XIth INTERNATIONAL ASTRONAUTICAL CONGRESS
STOCKHOLM 1960

XI. INTERNATIONALER ASTRONAUTISCHER KONGRESS

XIᵉ CONGRÈS INTERNATIONAL D'ASTRONAUTIQUE

PROCEEDINGS

VOL. II
SMALL SOUNDING ROCKETS SYMPOSIUM

EDITED BY
ÅKE HJERTSTRAND

PUBLISHED BY
THE ORGANIZING COMMITTEE OF THE CONGRESS

WITH 74 FIGURES

WIEN SPRINGER-VERLAG 1961

Editorial Committee

Åke Hjertstrand, M.Sc.

Carl Reuterswärd, Ph.D.

Arne Lindberg, Mr.

Acknowledgements:
The Small Sounding Rockets Symposium was included in the program of
the XIth International Astronautical Congress as an additional special meeting.
The Congress Committee is most grateful to acknowledge a generous contribution from
the UNESCO without which it would not have been possible to arrange for
that meeting and these proceedings.

ISBN 978-3-642-52020-4   ISBN 978-3-642-52037-2 (eBook)
DOI 10.1007/978-3-642-52037-2

AB AETATRYCK, Ahlén & Akerlunds tryckerier, Stockholm 1961
Softcover reprint of the hardcover 1st edition 1961

IV

Contents

List of Authors

Numbers refer to page

Developments of Project Kappa in 1959—60

Hideo Itokawa

Institute of Industrial Science of University of Tokyo, Chiba City, Japan

Project Kappa was a project to make an inexpensive, easy to handle, two-stage solid propellant, sounding rocket. Developments in 1959—60 could be classified into three fields. The first program was the refinements and extended actual service use of Kappa-6, which had been developed for IGY-1957/58. A newly developed propellant of higher specific impulse is used for the new Kappa-6, designated as Kappa-6H. Altitude performance at 80 degrees launching angle is increased from 60 km to 80 km, with the same payload.

The second program was to design and build a two-stage and a three-stage solid propellant rocket with the higher performance. The requirements for the two-stage rocket was a peak altitude performance of 200 km with 30 kg payload, and for the three-stage was 400 km peak altitude with 8 kg payload at 80 degrees launching angle. This program has primarily been developed from the requirements of the ionospheric research and the measurements of high energy particles. Along this line of the program, a new booster was developed. Combining with the rocket motors which had been made already for Kappa-6, a new two-stage rocket, Kappa-8, and a three-stage rocket, Kappa-9, were designed. A single-stage rocket with this new booster alone was also designed, and its flight test was conducted in November 1959. In March 1960, Kappa-8 with dummy second-stage rocket was fired, with the satisfactory results. Two Kappa-8 rockets were scheduled for flight in July 1960. The peak altitude attained by flight tests of Kappa-8 was 182 km with 35 kg payload at 78 degrees launching angle. The test flight of Kappa-9 is scheduled in January 1961, and it is supposed to climb 300 to 400 km at 80 degrees launching.

The third program in Project Kappa is the effort to cut down the cost of sounding rockets. A single stage rocket with glassfiber-reinforced plastic hardware has been developed under the designation of Pie. This third program is also being extended to the development of an inexpensive orbiting method, in connection with the application of a spherical motor.

Introduction

For IGY-1957/58, the Sounding Rocket Team of University of Tokyo had developed Kappa-6. Kappa-6 served for the rocket sounding for the observation of the temperature, wind, cosmic rays, pressure, and solar radiation. Further developments and the summary of its use in IGY will be reported in the next chapter.

At the end of 1958 a new booster was considered. This new booster was to be attached to Kappa-6, and thus make a three-stage rocket. For this purpose, two types of boosters were considered; the first type was a small auxiliary booster which would raise Kappa-6 to a little bit higher altitude, say to 80 or 100 km, but the principal idea was to increase the launching velocity through which the flight trajectory would be improved for the side wind dispersion. The burning duration of this small booster was designed to be 60 milli-seconds so that the burning will terminate on the launcher. Actually, this kind of small auxiliary booster was made and tested, however not used for actual flight test.

1

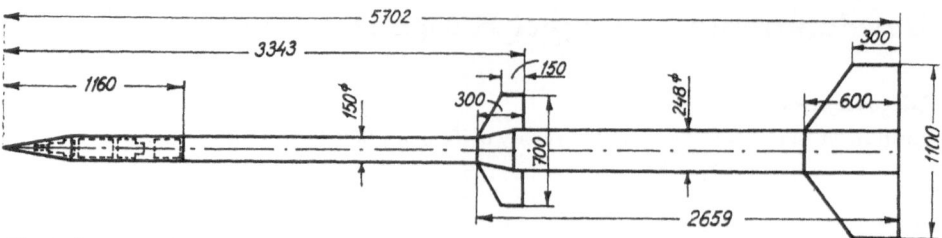

Fig. 1. Kappa-6 (Ø 150/248).

The second idea was to design a larger booster, which will make a complete new three-stage rocket.

The final dimensions of this new booster were: diameter 428 mm, length 6.3 m, and total weight 1.10 ton.

This new rocket motor can be used as a single-stage rocket, and can be used as a booster for two-stage rocket combined with any of the motors developed already for Kappa-6, and can also make a three-stage rocket. However, as a vehicle, special considerations are necessary for aerodynamic heating, stability, and aeroelasticity, because any of the rockets designed for Kappa-6 were made for lower speeds than if they had been designed as the second- or third-stage of a three-stage rocket.

Nose cones, walls, and tail fins would be subjected to a much higher kinetic heating in the case of the three-stage rocket, and some modifications in structure are necessary.

Fig. 2. Kappa-6 on the launcher.

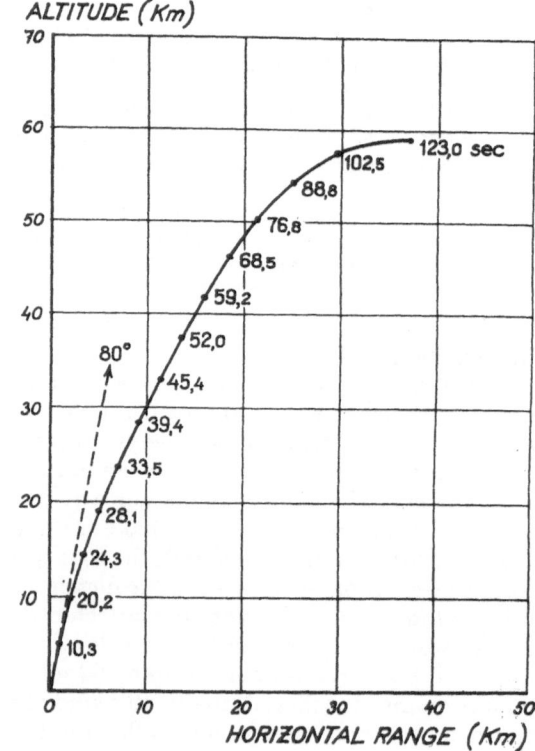

Fig. 3. A Kappa-6 trajectory by radar tracking data.

TABLE I. List of Firings of Kappa-6

	Date of Firing	Rocket	Angle of Launcher	Altitude (km)	Flight Duration (sec)	Purpose, Instruments
1	16 Jun. 1958	K-6-1	75°			Flight Test
2	20 Jun. 1958	K-6-2	78°	40—50	200	Flight Test
3	24 Jun. 1958	K-6-TW-1	78°	25	21	Temperature and Wind
4	30 Jun. 1958	K-6-TW-2	75°	45	75	Temperature and Wind
5	12 Sept. 1958	K-6-3	78°	40	208	Flight Test
6	14 Sept. 1958	K-6-4	78°	40	207	Flight Test
7	25 Sept. 1958	K-6-RS-1	78°	40—50	140	Solar Radiation
8	25 Sept. 1958	K-6-TW-3	78°	50	100	Temperature and Wind
9	26 Sept. 1958	K-6-TW-4	78°	60	100	Temperature and Wind
10	28 Sept. 1958	K-6-CP-1	78°	36	205	Pressure and Cosmic Rays
11	29 Nov. 1958	K-6-RS-2	78°	40	130	Solar Radiation
12	30 Nov. 1958	K-6-CP-2	78°	49	230	Pressure and Cosmic Rays
13	23 Dec. 1958	K-6-TW-5	80°	60	120	Temperature and Wind
14	17 Mar. 1959	K-6-RS-3	80°	56	240	Solar Radiation
15	18 Mar. 1959	K-6-TW-6	80°	50	104	Temperature and Wind
16	19 Mar. 1959	K-6-RS-4	78°	50	215	Solar Radiation
17	20 Mar. 1959	K-6-TW-7	80°	50	104	Temperature and Wind

Assuming that these necessary modification will be done, the new single-stage rocket will be called Kappa-7, and the two-stage rocket formed by combining the 428 mm booster with the first-stage rocket of Kappa-6 is designated as Kappa-8. The three-stage rocket which is the combination of Kappa-6 and the new 428 mm motor is designated as Kappa-9.

The first flight test of Kappa-7 was conducted in November 1959, and the Kappa-8 with a dummy second-stage rocket was fired in March 1960. The first real two-stage Kappa-8 was launched in July 1960, the firing of the three stage Kappa-9 is scheduled for January 1961.

Remarks on Kappa-6

Before proceeding into the new Kappa-series, the basic Kappa-6 is reviewed here. Kappa-6 is a two-stage, solid propellant, sounding rocket, capable of climbing to an altitude of over 60 km with a payload of 5 to 10 kg at the 80 degrees launching angle. The diameter of the first-stage rocket is 245 mm, and its length is 2.66 m. Total launching weight of Kappa-6 is approximately 360 kg, and total length lies in between 5.4 and 6.0 m.

Fig. 1 shows a typical layout of Kappa-6, and a picture of the rocket on its launcher is shown in Fig. 2.

Fig. 3 shows a flight trajectory of one of Kappa-6 rockets observed by radar tracking. Special attention has been paid in developing Kappa-6 to inexpensiveness, operation simplicity, and safety in handling. Only solid propellants have been considered in Kappa-series, and Project Kappa was and is a completely civilian project from its basic research to the final firing operation. Seventeen Kappa-6 rockets were fired up to March 1959. Table I is the list of Kappa-6 firings.

The standard instrumentation for test flight were the accelerometer, decelerometer, skin temperature gauge, and strain gauge. For the observation of the temperature and wind of the upper atmosphere, the grenade method was adopted. A Geiger counter was used to study cosmic ray intensity variation with altitude, and a Pirani gauge was installed for the pressure measurement. For the solar radiation observation, a solar spectrograph was installed in the nose cone and recovered from the ocean with the aid of the parachute and float.

In 1960, two Kappa-6 flights are scheduled, and some modifications are being considered. Fig. 4 shows the modified Kappa-6 which is called Kappa-6H.

The principal difference between the nominal Kappa-6 and Kappa-6H lies in its longer booster with the same diameter, and the application of a new propellant of higher specific impulse. This will improve the altitude performance from 60 to 80 km at 80 degrees launching, and with vertical launching it will climb up to the altitude of 90 to 100 km. Two Kappa-6Hs are scheduled for firing in September 1960 for the measurement of temperature and wind, participating in the World Rocket Week coordinated by COSPAR.

TABLE II. The Telemetering System of Kappa-6

Modulation	FM-FM
Carrier frequency	225 Mc
Subcarrier frequency	2.3, 3.0, 3.9, 5.4, and 7.35 kc
Number of channels	five
Rf output power	1 watt
Weight of rocket borne transmitter	2.8 kg
Size of transmitter	140 mm/235 mm

For radar tracking, three systems are used; the automatic range and direction tracking radar (Modified GMD-1), four sets of range finders, and DOVAP system. For range and direction finders the frequency of 1680 Mc is used, and for the DOVAP system the frequency is 40—80 Mc. For both tracking and DOVAP system, rocket borne transmitters are being used (beacon system). Optical tracking systems of various kinds developed for Kappa-6 are being used. The Akita Rocket Range of University of Tokyo is at 39° 34′ 12″ N, 140° 03′ 35″ E.

The New Booster, 420B

TABLE III. Requirements for a New Booster

(1) It must be attached to Kappa-6, with the adequate stability and aeroelasticity requirement.

(2) It must boost Kappa-6, so that the final stage rocket attains the altitude of 100—200 km and enables the sounding of ionosphere and cosmic rays.

(3) The launching velocity by this booster should not be so low that sidewind dispersion affects the trajectory remarkably.

(4) The motor should be fabricated by welded sheet plate of high tensile steel.

In order to meet with the first requirements, a theoretical study of the optimization of three-stage solid propellant rockets was carried out [3]. Varying the diameter of booster from 250 mm to 750 mm, the following data were calculated:

Center of gravity location	Initial acceleration
Mass ratio	Loading density
Length-diameter ratio	Propellant weight
Thrust	Motor weight

Developments of Project Kappa in 1959—60

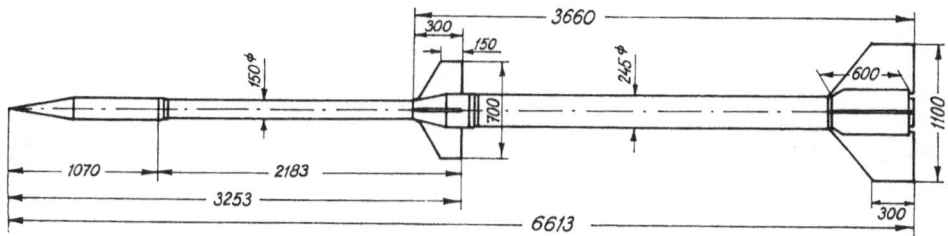

Fig. 4. Kappa-6 H (∅ 150/245).

Performance calculations were carried out for the case of the following three diameters: 400 mm, 450 mm and 500 mm.

The results were as follows:

TABLE IV. *Performance Calculations for Different Kappa-Configurations*

Designation	Total weight (kg)	Altitude by launching angle of	
		90 degrees (km)	80 degrees (km)
Kappa-7 (400)	900	60	50
Kappa-8 (400)	1160	280	230
Kappa-9 (400)	1230	490	400
Kappa-7 (450)	1260	90	80
Kappa-8 (450)	1530	310	260
Kappa-9 (450)	1600	530	440
Kappa-7 (500)	1900	120	100
Kappa-8 (500)	1970	350	290
Kappa-9 (500)	2040	600	500

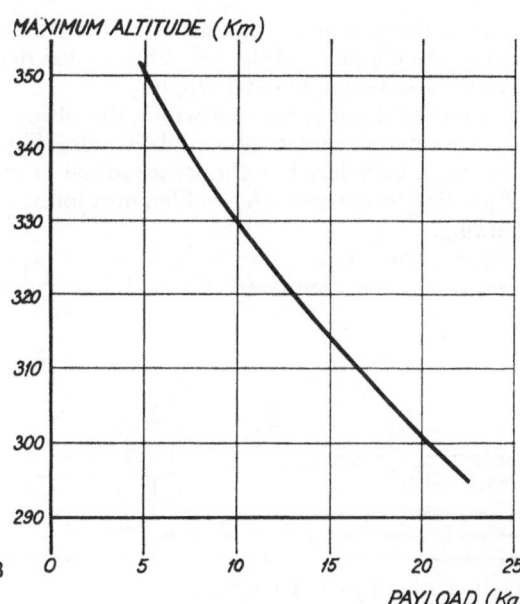

Fig. 5. Altitude versus payload of Kappa-8 (∅ 248/400 at 90° launching angle.

5

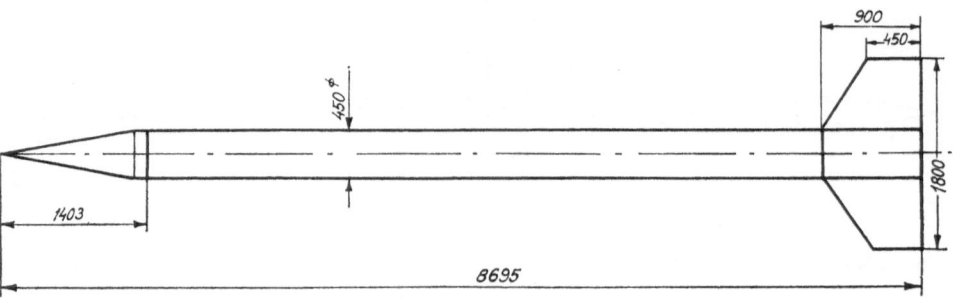

Fig. 6. Kappa-7 (Ø 450).

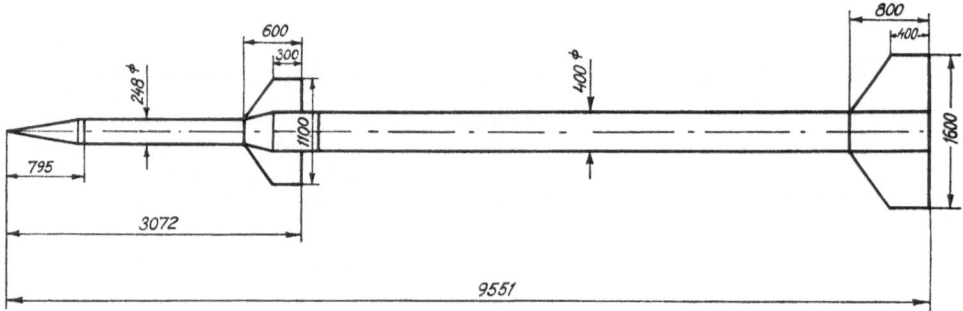

Fig. 7. Kappa-8 (Ø 248/400).

In the computation of the performance, the specific impulse was kept constant, at 200 sec, for all stages for the sake of simplicity.

The effect of payload increase was studied, and for a case of the 400 mm diameter series, the results are shown in Fig. 5.

The general layout of the rockets designated above are illustrated in Fig. 6, 7, and 8.

After the extensive evaluation of these results together with the other requirements, the diameter of the new booster was decided to be 420 mm finally. This new booster was designated as 420B, Fig. 9.

High tensile steel plate of which the ultimate tensile strength is over 85 kg/mm² was considered for the motor fabrication. The sheet plates were rolled and welded to form a cylinder. For the longitudinal assembling, the following three methods of jointing were tried: (A) welded butt joint; (B) screwed jointing; and (C) riveted jointing.

These three types were actually made experimentally in full scale, using actual steel plate, and compared. From the view point of the dimensional accuracy, de-

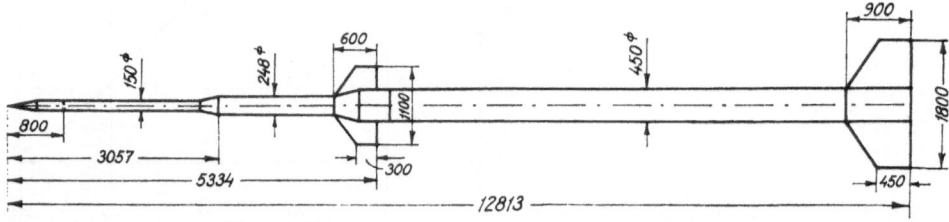

Fig. 8. Kappa-9 (Ø 150/248/450).

Fig. 9. 420B motor.

formation allowances, and aerodynamic cleanness, finally the type of (A) was chosen for 420B.

In order to obtain a cylindrical shell of high dimensional accuracy, two methods of thickness determination were studied. In the first case, thicker plate of the thickness of about 7 mm was rolled and welded to the cylindrical shell, and then machined at both inside and outside surfaces to the ultimate thickness of 2.8 mm.

In the second case, plate was rolled to the thickness of 2.8 mm in advance, and then rolled up to cylindrical shell directly. The former was finally chosen.

The yield point of the material is 84 to 90 kg/mm² with elongation of 18—20%. At the welded point of the chamber the tensile strength is higher than 85 kg/mm² and Charpy test value for impact is 6 kgm/cm² at 0 °C.

Kappa-7

As already stated, Kappa-7 is a single-stage rocket with the 420B motor. Fig. 10 shows its general layout.

Fig. 10. Kappa-7 rocket.

Kappa-7 was designed for two purposes. First, it serves for the flight test of a new booster 420B, and second, it could serve as a sounding rocket of approximately the same altitude performance as Kappa-6, but with a higher payload of say 40 to 60 kg.

However, the final weight estimation and flight performance showed that Kappa-7 has a comparatively low altitude performance as a sounding rocket. From this reason, instead of designing the so called simultaneous multi-observation sounding rocket, only one Kappa-7 was made and flight tested for the purpose of flight check of 420B.

The dimensions of Kappa-7 are:

TABLE V. The Dimensions of Kappa-7

Length	6720 mm
Diameter	425 mm
Total Weight	1200 kg
Maximum Altitude at 80° Launching Angle	50 km
Total Flight Time	220 sec
Horizontal Distance of Point of Impact	50 km
Payload	30 kg

The instrumentation installed in Kappa-7 were:

TABLE VI. Instrumentation Installed in Kappa-7

Accelerometers along two axis	
Decelerometer along thrust axis	
Vibration meter	
Temperature gauge at nose-cone and on tail fin	
Strain gauges on tail fins	
Telemeter transmitter	225 Mc FM-FM five channels
Radar transponder	1680 Mc
DOVAP transponder	39 Mc—78 Mc

A new electronic device of the parametric amplifier was adopted in the system of DOVAP, and satisfactory results were obtained for future low noise space electronic system.

A new launcher of movable type was prepared for Kappa-7. However, this new launcher needed a modification for the further use. The exhaust gas flame of the motor climbed back to the direction of nose at a certain time interval during launching stage. This phenomenon was shown by the high speed camera data of Kappa-7. A slight modification was made after the launching of Kappa-7, which was conducted on November 18, 1959.

This phenomenon of back-flaming never appeared in the case of launching of Kappa-8D.

Kappa-8D

Preceding the firing of Kappa 8, flight test of Kappa-8 with a dummy second-stage rocket was planned, and this was called Kappa-8D.

The dimensions of Kappa-8D are approximately the same as Kappa-8.

Fig. 11. Kappa-7 on the launcher.

TABLE VII. The Dimensions of Kappa-8D

Total length	10.12 m	
Length of the second-stage rocket	4.40 m	
Length of the first-stage rocket	6.34 m	
Diameter of the second-stage rocket	250	mm
Diameter of the first-stage rocket	428	mm
Total weight	1480	kg
Weight of the second-stage rocket	292	kg

The second-stage rocket was never detached from the first stage during the whole flight time, therefore Kappa-8D was actually a single-stage rocket. The purpose of

Fig. 12. Kappa-8D on the launcher.

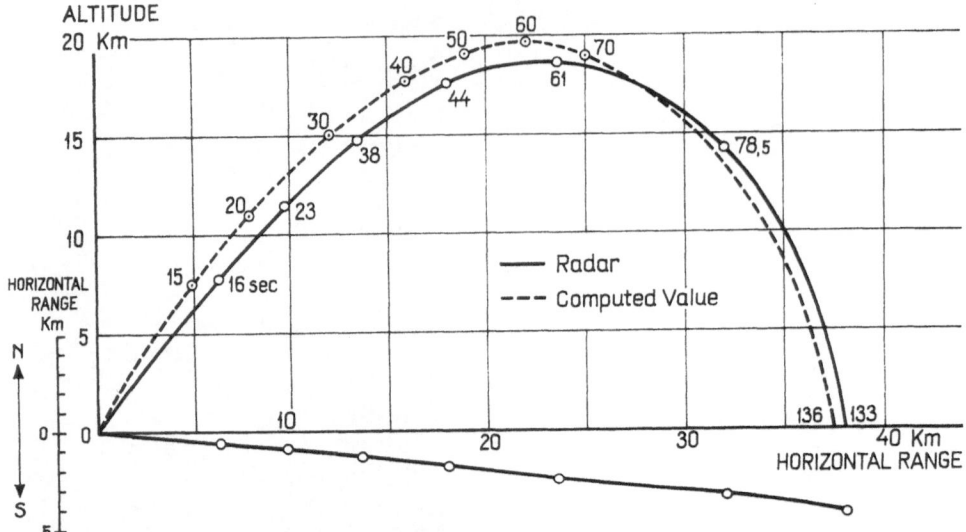

Fig. 13. Computed and radar-tracked trajectory of Kappa-8D.

Kappa-8D was to study the stability characteristics with the second stage rocket. The instrumentation installed in Kappa-8D are shown in Table VIII:

TABLE VIII. Instrumentation Installed in Kappa-8D

Accelerometer along thrust axis
Accelerometer along transverse axis
Decelerometer along thrust axis
Temperature gauge on nose cone, and tail fins
Strain gauges on tail fins
Telemeter transmitter
Radar transponder
DOVAP transponder

The antenna for the rocket borne radar transponder was a horn type antenna attached to the body of the second-stage rocket, and for telemetering, a whip antenna was set on the trailing edge of the fin of the second-stage rocket.

For the DOVAP system, a whip antenna was attached to the trailing edge of the fin of the first-stage rocket. Flight test of Kappa-8D was carried out on March 28, 1960 at the Akita Rocket Range of University of Tokyo. The launching angle was 65°.

The actual flight trajectory observed by radar system coincided remarkably with the calculated trajectory, Fig. 13. The aerodynamic stability was proved to be satisfactory, too. The total flight time by telemetering system was 133 sec. In Fig. 14 telemetered data during the flight are shown.

Kappa-8

A further step forward in the design of Kappa-8 was made by the success of the flight test of Kappa-8D. Kappa-8 is a two-stage sounding rocket, which is a combination of Kappa-7 and the first stage rocket of Kappa-6.

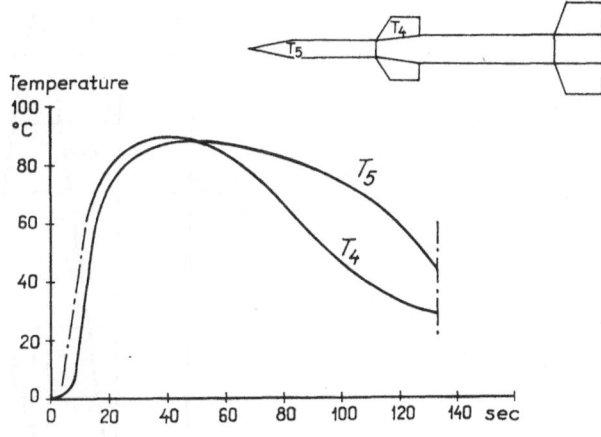

Fig. 14. Skin-temperatures measured during flight of Kappa-8D.

TABLE IX. The Dimensions of Kappa-8

Total length of two-stage configuration	10.032 m
length of second-stage rocket	4.393 m
length of booster rocket	6.264 m
Total weight of Kappa-8	1476 kg
weight of second-stage rocket	306 kg
weight of booster rocket	1170 kg
Max. diameter of booster rocket	439 mm
diameter of second-stage rocket	259 mm
Semi-apex angle of nose cone	10 deg
Fin configuration, four fins	
fin span of booster	1.180 m
fin span of second stage rocket	1.100 m
Payload	35 kg

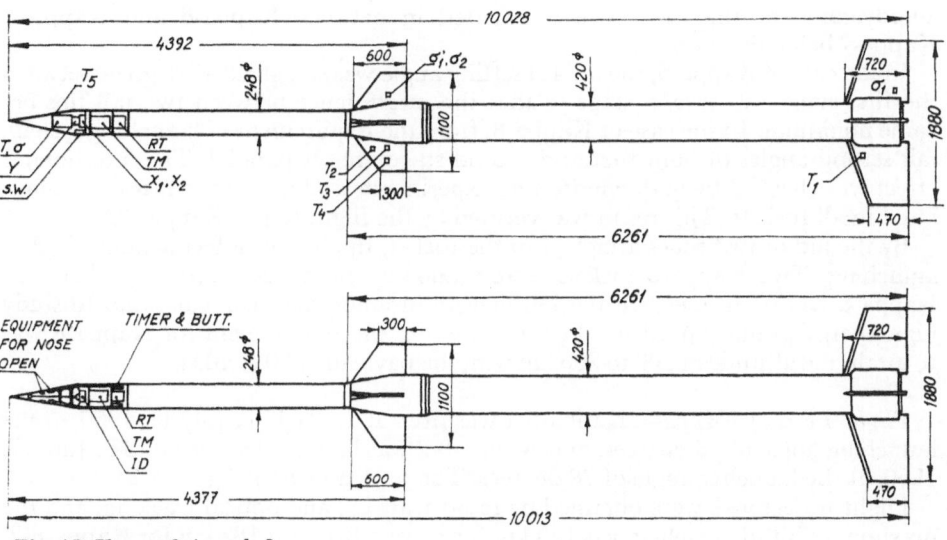

Fig. 15. Kappa-8-1, and -2.

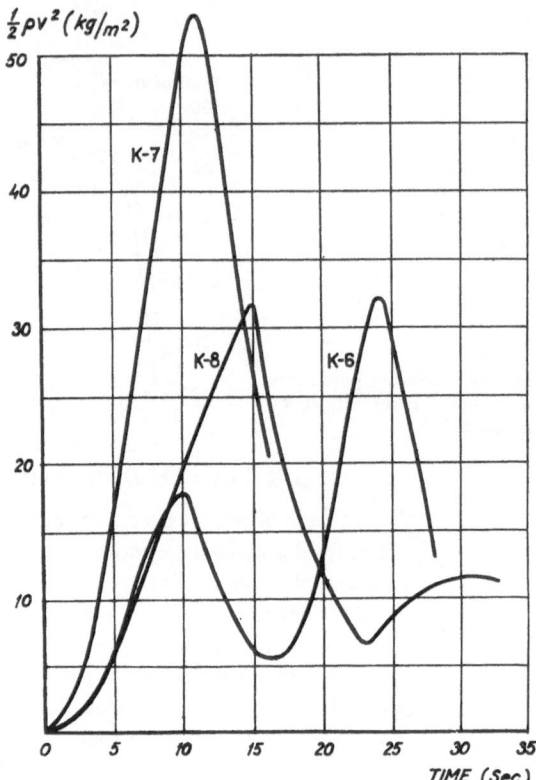

Fig. 16. ϱv^2-time-diagram of Kappa-6, -7, and -8.

In Fig. 16, and 17, ϱv^2, and $v^{2.5}\varrho^{0.5}$ are shown for the each case of Kappa-6, Kappa-7, Kappa-8D, and Kappa-8. These are factors which influence the aerodynamic heating. From these diagrams, it is easily seen that from the view of the aerodynamic heating, Kappa-8 is situated in between Kappa-6 and Kappa-7, Kappa-7 being the severest.

In the case of Kappa-6, the tail fin setting angle was staggered 45 degrees between the first stage and second stage so that the interference between two tail fins became minimum. In the case of Kappa-8, from the requirement of launching system, tail setting angles of both first and second stage are set parallel. The interference effect was checked by both windtunnel experiments and actual flight tests of simulated small rockets. This result was verified by the flight test of Kappa-8D.

By the aid of two shoes attached to the rocket, tipping off effect is minimized in launching. Two Kappa-8 rockets were made and flight tested, and another four Kappa-8 will be to measure the ion density in ionosphere E-region, and to study high energy primary particles. For this mission, the requirement for Kappa-8 was to attain the altitude of 100 to 200 km with the payload of 30 to 40 kg.

Flight Tests of Kappa-8. Kappa-8-1 was fired at 13:24 JST, July 11, 1960, at the launching angle of 73 degrees, and Kappa-8-2 was launched at 13:11 JST, July 17, 1960, at the launcher angle of 78 degrees. The payload was 35 kg.

Flight trajectories were obtained by radar tracking and optical tracking, and the maximum altitude reached was 150 km for Kappa-8-1, and 182 km for Kappa-8-2.

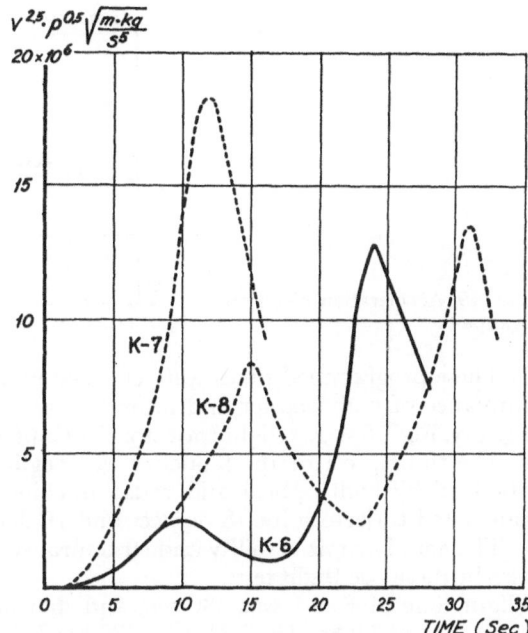

Fig. 17. $v^{2.5}\varrho^{0.5}$-time-diagram of Kappa-6, -7, and -8.

In both cases the wind on the ground was 3 m/sec against launching direction. The followings are the data for the thrust program:

TABLE X. *Thrust Time Program*

Time of b.o. of booster	15 sec
Separation time of booster from second rocket	17 sec
Ignition time of second-stage rocket	26 sec (Kappa-8-1)
	30 sec (Kappa-8-2)
Time of b.o. of second stage rocket	35 sec (Kappa-8-1)
	39 sec (Kappa-8-2)

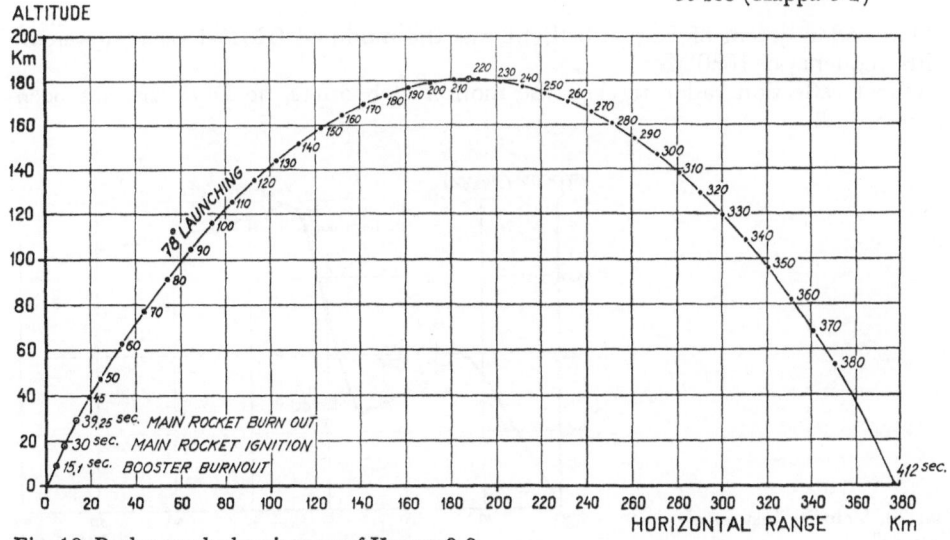

Fig. 18. Radar-tracked trajectory of Kappa-8-2.

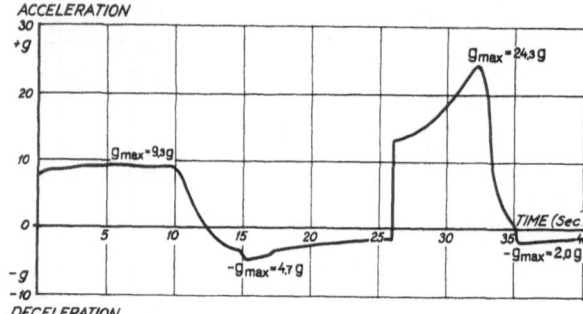

Fig. 19. Acceleration history of Kappa-8.

These programmed times were checked by telemetering system, and engine performance of both booster and second staged rocket have been proved to be satisfactory. Fig. 18 shows flight trajectory of Kappa-8-2.

The tipping off of the launcher was negligibly small, and the flight trajectory obtained by both optical and radar tracking systems coincided closely with the calculated trajectory for 73 degrees and 78 degrees without wind.

The aerodynamic stability and structural strength were proved to be satisfactory also by these two flight tests.

Flight time of K-8-1 was 330 sec, and that of K-8-2 was 412 sec and horizontal ranges were 360 km (K-8-1), and 372 km (K-8-2).

Fig. 19 and Fig. 20 show acceleration and velocity history of Kappa-8, and maximum speed is approximately 2000 m/sec, and maximum acceleration 24 g.

The skin temperature at nose-cone measured by telemetering system during flight of K-8-1 was 260°C. These temperatures coincide well with the computed aerodynamic heating data, though the actual measured value were little under the estimated value. An ablating material was coated on the surface of nose-cone, and tail fin surfaces.

The coasting time of K-8-2 was 4 sec longer than that of K-8-1, and this prolonged coasting time aimed to decrease the skin temperature rise.

The telemetering system of K-8 was the same as that of Kappa-6.

The radar system of Kappa-8 flight was the modified GMD-1 beacon system, with frequency of 1680 Mc.

Direct reflection radar tracked the motion of boosters, no DOVAP, has been used for K-8.

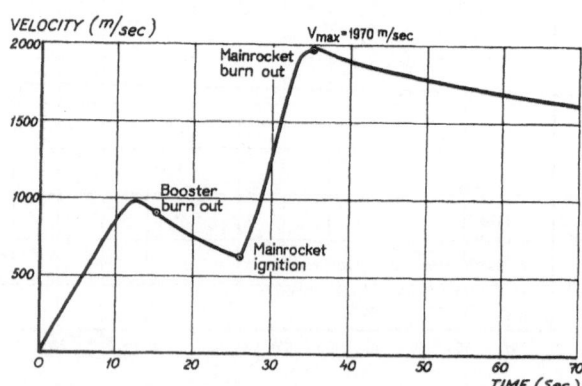

Fig. 20. Velocity history of Kappa-8.

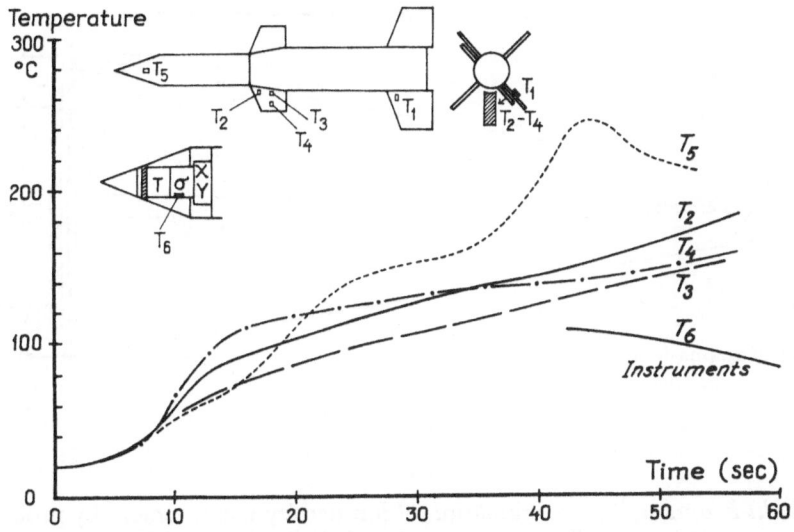

Fig. 21. Temperatures measured during flight of Kappa-8.

TABLE XI. *Radar System Data*

Frequency	1680 Mc	Weight	2.9 kg
Repetition of pulse	500 cps	Outside diameter	147 mm
Pulse width	1 sec	Length	156 mm

Whip antenna was used for telemetering, and notch antenna for radar. Both antennas were fitted at the trailing edge of tail fin of the second stage rocket with the particular consideration for aerodynamical heating.

TABLE XII. *Flight Test Instrumentation of Kappa-8*

Accelerometer along thrust axis
Decelerometer along thrust axis
Accelerometer perpendicular to thrust axis
Vibration meter along thrust axis
Temperature gauges at nose cone, tail fin surfaces,
    instrumentation portions
Strain gauges on the tail fins

In Kappa-8-2, a probe type ion density instrument was installed for preliminary experiment of the ionosphere observations.

The nose cone was to be split into two pieces by a programmed timer, and the ion probe exposed to external space. Particular consideration was paid to the design of the probe to minimize the photoemission contamination effect.

A cage type configuration was adopted for this purpose, and proved by ground based experiment to be satisfactory. However, during the flight of K-8-2, because of the malfunction of the splitting mechanism of the nose cone, the actual ion density measurement in the E-region failed.

15

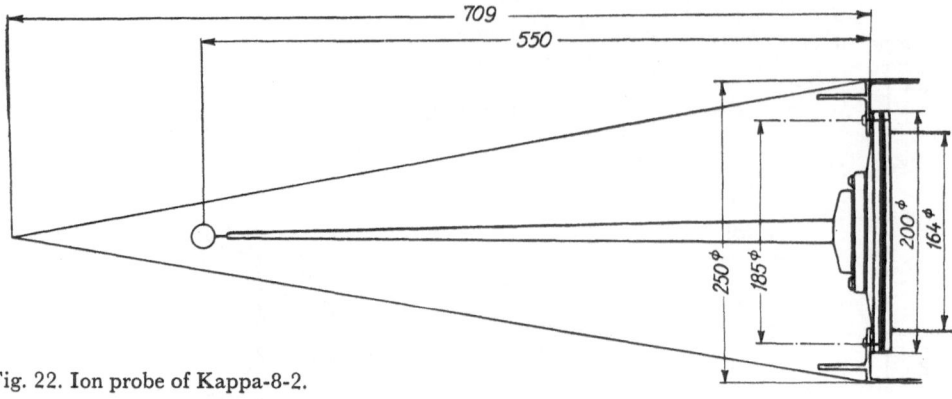

Fig. 22. Ion probe of Kappa-8-2.

Further Flight Program. The continuation of ion density measurement by probe method will be carried out in September, during COSPAR World Rocket Week (Sept. 16—22, 1960). Another two Kappa-8 flights are scheduled in 1960 fiscal year, one for ionosphere observation for mid-winter, and one for mid-spring.

Kappa-8 will serve for the study of the ionosphere, by the probe method for ion density and resonance impedance method for electron density measurement.

Also Kappa-8 will install the several instrumentations for the study of high energy particles, cosmic rays, together with ionosphere instrumentations.

Project of Kappa-9

Kappa-9 is a three staged sounding rocket which is a combination of Kappa-6 and Kappa-7, Fig. 8. Total length is 12.3 m, and total weight is 1.6 ton. The peak altitude at vertical launching will be 420 km, and that of 80 degree launching will be roughly 360 km.

Another project is being studied under the designation of Pie, which aims to develop particularly inexpensive sounding rockets. Glassfibre-reinforced plastic chamber is experimented with a long duration burning propellant.

Research and Development of Spherical Motors

Spherical shaped rocket motor has been developed since 1959 in connection with an inexpensive orbiting technique.

A spherical motor has two distinguished advantages over usual cylindrical rocket motors:

(1) *easy to be installed in the nose cone of a booster, in gimbal-mounted condition, given spinning pre-flight.*
(2) *high mass ratio.*

Fig. 23 shows its function as applied to sounding rocket and orbiting a small satellite. By applying spherical motor, it may be possible to eliminate guidance and control system from orbiting technology. High mass ratio of the order of 15 can be realized by spherical motor [4]. Both steel and glassfibre-reinforced plastic hardware have been developed, and static firings were carried out for three-dimensional star type solid propellant. Flight tests of spherical motors are scheduled for August 1960.

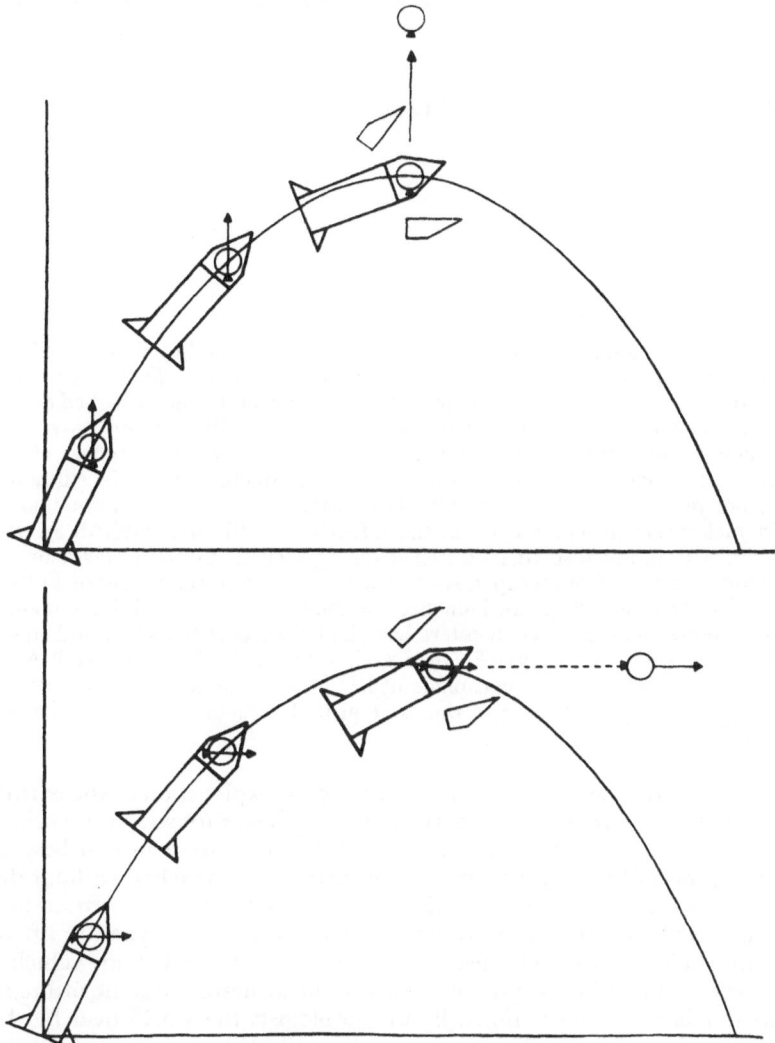

Fig. 23. Spherical motor for sounding rocket and satellite.

References

[1] *Kimura, K., Takei, T.,* and *Ogawa, K.,* Plastic Hardwares for Rocket Motors, Proceedings of the First International Symposium on Rockets and Astronautics, Tokyo, 1959, p. 107 (Japanese Rocket Society).

[2] *Itokawa, H.,* Project Kappa. Proceedings of the First International Symposium on Rockets and Astronautics, Tokyo, 1959, p. 188 (Japanese Rocket Society).

[3] *Kakimi, T.,* On the Optimum Design of Three-Stage Solid Propellant Rockets with Special Reference to their First-Stage Rockets. Proceedings of the First International Symposium on Rockets and Astronautics, Tokyo, 1959, p. 94 (Japanese Rocket Society).

[4] *Akiba, P., Itokawa, H.,* On the Spherical Solid Rocket Motor. Paper presented at the Second International Symposium on Rockets and Astronautics, Tokyo, 1960 (Japanese Rocket Society).

Small Rockets for Research and Weather Observation

Willard S. Houston, Jr.

Office of Naval Research, Branch Office, London, England

The US Navy has long been active in the field of small sounding rockets and has participated in many development programs. After the Viking, which was a large liquid fuel rocket over a meter in diameter, 13 meters long and which weighed nearly 7500 kg, the general trend has been toward smaller more efficient and lower cost rockets. These systems are used both for the investigation of the atmosphere of the earth and to place instruments above the earth's atmosphere in order to study the sun and other planets. The Aerobee and Aerobee-Hi represent the continuation of the liquid fuel rocket development and the DEACON, ASP and ARCAS are the result of efforts to produce small solid fuel sounding rockets. Efforts to make use of rocket motors borrowed from weapon systems have resulted in the HASP or LOKI-DART and MIRAL or 2.75 in. air launched sounding rockets. Combining sounding rockets with larger weapon type boosters has yielded such well known and much used research vehicles as the NIKE-DEACON, NIKE-CAJUN and NIKE-ASP. The latter system (NIKE-ASP) is capable of lifting scientific payloads of 1/10 the weight of those carried by Viking to the same altitude reached by the Viking but at 1/50th of the cost.

The type of research that we are referring to here is the exploration of the earth's atmosphere with the aid of rockets which are used as vehicles to carry the various sensors to the desired altitude from which a particular measurement can best be made. As we know, most of the atmosphere of the earth can be reached without the use of rockets. Indeed, approximately 99 % of the gaseous envelope surrounding this planet is accessible directly from aircraft which require aerodynamic lift to enable them to fly. This includes both heavier than air and lighter than air vehicles. Probably the most common aircraft used for study of the atmosphere is the ordinary extensible rubber balloon which is flown by meteorologists the world over by the hundreds of thousands each year. Treating the balloon itself as a sensor, winds are obtained by tracking it with optical or electronic ground equipment. By adding a small radio transmitter and suitable measuring devices pressure, temperature, and humidity can also be obtained along the balloon's trajectory as it ascends. This information is transmitted back and recorded on the ground. Parameters such as refractive index, ozone concentration, solar and terrestial radiation, concentration of airborne particle matter and others are also determined in this fashion. On a routine basis soundings up to about 80 000 ft or 25 km are obtained. By using larger rubber balloons and giving them special care this altitude can be increased to approximately 110 000 ft or about 34 km. Balloons made of very light weight non-extensible plastics have gone as high as 147 000 ft or 45 km. For routine measurements however the balloon ceiling can be considered about 100 000 ft or 30 km. Above this altitude is the region of the rockets. Much above this altitude, say, from about 150 to 200 miles or approximately 300 km on up, satellites may be used effectively for certain purposes.

Small Rockets for Research and Weather Observation

Between 30 km and 300 km the rocket is king. It is also this region of the atmosphere that is particularly interesting as it is the transition zone between the air of nearly uniform composition which behaves like an ideal gas, and the region where densities are so low, and mean free paths are so large that the air can no longer be treated as a fluid. A large portion of the ionosphere also falls in this zone and major changes in the chemical action take place as the sun's radiation is selectively absorbed by the various atmospheric constituents. In this, the rocket region, observation time is short and fast response instruments must be used. In the balloon belt below observation times are measured in hours and days, and in the satellite space above observation times are measured in months and years. With rockets, however, passage through the zone of interest takes only seconds or at most minutes. Because of this very short dwell time at the place of observation, and because observations usually are only possible along a single nearly vertical path many rockets are required to obtain a description of this portion of the atmosphere and to determine time and space variations in its structure, composition, and electrical and electromagnetical properties. A few observations taken at different latitudes at different times of day and in different seasons can serve only to create a very rough picture of what is occurring in this fascinating region.

In order to obtain more observations with better geographical distribution and more nearly synoptic in time, low cost easily launched rockets are needed. Generally speaking, the requirements of low cost and ease of handling also implies small size. Reducing the cost of individual rocket flights also permits more groups to participate in research programs where rockets are needed and by broadening the base for high altitude research it will, perhaps, hasten the time when man can make full use of the unique properties found there. Since the beginning of the upper atmospheric research programs which were initiated with German V2 rockets in 1945, development of simpler and cheaper sounding rockets has been carried on by scientists of the US Navy. The Aerobee and Aerobee-Hi rockets developed by the Naval Research Laboratory permitted a much larger research program than would have been possible with the V2.

As a sounding rocket the V2 was large and relatively inefficient and possessed little flexibility with respect to payload distribution or payload weight. It was approximately 40 ft long and weighed nearly 15 ton at take off. The Aerobee measures approximately 25 ft in length and weighs only 1 ton including the booster. Somewhat paralleling the Aerobee development was a similar but less intensive program involving the Viking Rocket (35 ft long, about 4 ton take off weight). The Viking stood midway between the V2 and Aerobee in payload capacity and vehicle complexity. The attendant reduction in cost and increased facility in handling of the Aerobee rocket permitted more observations and made multiple launching sites economically feasible, within the financial structure of the research budgets. The introduction of the Aerobee caused a reduction in the size and weight of the scientific payloads which would be flown, but this was in some respects a blessing in as much as each flight represented fewer experiments than had previously been flown on the larger rockets and less risk was associated with each launching and fewer compromises were required in instrumentation. Improvements on the Aerobee rocket made in 1952 resulted in the Aerobee-Hi which is capable of raising 75 kg to an altitude of about 250 km. Although smaller and simpler than the V2 or the Viking the Aerobees are still large rockets and represent a sizable investment for each shot and a very considerable investment to prepare a launching site for their use.

About the time that the Aerobee-Hi rocket was being developed another approach to the problem of costs was taken by a Navy sponsored group at the State

University of Iowa. The Iowa approach was to lift small rockets to altitudes of 70 000 or 80 000 ft (21 to 24 km) by means of helium filled balloons and then to launch them by radio signal from this elevated position. This technique avoided the rapid loss of velocity experienced by a high speed projectile passing through the dense air near sea level and was equivalent to providing a booster stage of approximately five to ten times the total impulse of that of the small rocket. A number of flights using this technique which was named Rockoon were made during the period 1952 to 1958. The most marked advantage of such a system is the extreme low cost relative to the altitude reached by the scientific payload. Other advantages of small size, light weight, ease of handling and reduced aerodynamic heating of the rocket are also present but to a lesser degree because of the balloon. One major limitation of the system is that during the long uncontrolled ascent of the balloon the rocket is often carried many miles from the release point and the flight of the rocket to its impact point is only predictable within very wide limits. Thus, launching near inhabited areas are usually not feasible. Also conditions are not always suitable for launching balloons of the size needed.

The rockets used in the Rockoon work were the DEACON and LOKI both of which were developed initially for use other than for atmospheric research. A number of solid propellent rockets not originally designed for the purpose have been used for exploration of the upper atmosphere. Among those worthy of special mention are the NIKE-CAJUN and NIKE-DEACON combinations which were used extensively during the IGY. These vehicles are representative of the application of weapon propulsion units to the task of providing sounding rockets for atmospheric research. Because most of these rockets used solid propellents a new degree of handling ease and simplicity of launching was brought into the research program. Also because many of these units were surplus they were available at very little cost to the research program. A typical operation involving the launching of a NIKE-CAJUN rocket requires only two to three hours for preparation of the vehicle and firing. Also the launching equipment is smaller, more portable, and much less expensive than that required by the larger liquid rockets. Together with these advantages, however, came certain restrictions principally as a result of the difference between the desirable properties of a weapon and a sounding rocket. Weapons, generally speaking, should have high acceleration and high velocity in order to simplify launching and to minimize flight time and target motion. These requirements are not only not required in a sounding rocket but are, indeed, a handicap as the high acceleration places unwanted large strains and stresses on the instrumentation and the high velocity, particularly at low altitudes, is an uneconomical use of the available thrust and creates undesirably large aerodynamic heating factors. This latter problem often requires an amount of heat-shielding equivalent in weight to the useful payload.

Under certain conditions, however, the desirable properties of the weapon are needed also in the sounding rocket. An example of this is the Bureau of Naval Weapon's program. The HASP which stands for High Altitude Sounding Projectile utilizes the propulsion unit from the obsolete Army anti-aircraft system designated LOKI. This motor measures 76 mm in diameter and is 2 meters long. As modified for the HASP it gives very large accelerations (300 g) and high heating factors but is capable of being fired under nearly all weather conditions both from the barrel of a 5 in. naval gun and from a simple tube launcher, thus enabling it to be used for routine weather observations not only above balloon altitude but under conditions when balloons could not be launched. The instruments are housed in a dart (34 mm in diameter by 102 cm in length) which separates from the motor at burn-out and then coasts to peak. As currently used the HASP carries only a radar target either of dipoles or a parachute which is released at the peak of the trajectory and which travels with the wind during descent. By tracking the target, high altitude winds are obtained. A fully instrumented version of he HASP is under development by the Naval Ordinance Laboratory but represents a difficult problem in miniaturization and sensor capability.

Another use of a small obsolete weapon for meteorological observations has been the adaption of the 2.75 in. FFAR for use as an aircraft-launched sounding rocket. This motor is 70 mm in diameter and approximately 1 meter long. It is equipped with folding fins which permits launching from a tube only slightly larger than the motor. Although originally designed to be fired forward from an aircraft this vehicle has been successfully launched into a vertical trajectory from a horizontally flying aircraft. This is achieved by pointing the launcher at an angle of approximately 45 degrees aft of vertical and relying upon the weathercocking effect of the relative wind caused by the aircraft's motion to tip the rocket into vertical flight. The advantages of such a system in permitting high altitude observations from remote locations without the need for surface installations is obvious. Although limitations in both altitude capability and instrumentation load are imposed by the use of such a small rocket the technique has been demonstrated to be feasible and the altitude and load restrictions can be overcome by using larger rockets designed for this purpose.

As mentioned before the use of obsolete weapons as sounding rockets while suitable in many cases represents a compromise in performance characteristics which is not always acceptable. Therefore, in addition to adapting existing vehicles to research use and exploring their capabilities as vehicles for obtaining routine meteorological observations, the US Navy has also sponsored development work to produce rockets with the desirable capabilities found in the larger liquid propelled vehicles such as Aerobee and yet with the reliability and simplicity of the solid fueled motors.

One of the initial efforts in this direction was in the development of the ASP. This single stage solid propellent rocket was designed and built by the Cooper Development Corporation under contract to the US Navy Bureau of Ships. Development of the ASP was initiated and carried out as part of a program to obtain certain atmospheric observations during Operation Redwing but has been subsequently used for many research projects as a single stage, a first stage and a final stage motor in various sounding rocket configurations. The basic dimensions of the ASP are 165 mm in diameter and 3.65 meter in length. It is capable of transporting an 8300 cm^3 payload weighing approximately 11.5 kg to an altitude of 61 km. The propulsion unit is an internal burning solid propellent with a burning time of approximately 6 sec. The motor produces an average thrust of 2650 kg at sea level at normal temperature and achieves burn-out velocities in access of 1650 m/sec when launched from sea level. Although still a comparatively short burning time the ASP represents a marked improvement over earlier solid propellent rockets of similar size which have been used for sounding rockets with an attendant increased altitude capability. When used as a second stage in combination with a NIKE-AJAX booster, the ASP will carry a scientific payload of 20 kg to an altitude in excess of 300 km. The NIKE-ASP configuration has been used in a number of Navy sponsored research projects. Launched from a simple rail launcher which can be easily erected in remote sites and abord ship, a NIKE-ASP can be utilized for experiments where the rocket must be fired upon demand at the scientifically opportune moment to measure desired transient phenomena. An example of this was the Naval Research Laboratory's Solar Eclipse expedition conducted from shipboard in the fall of 1959 when six NIKE-ASP rockets were fired within ten minutes in order to make measurements during the various phases of the eclipse.

One of the most recent developments in the small sounding rocket field was made possible by the development of high performance fuels suitable for use in restricted burning motors. This end-burning technique permits the achievement of long burning times comparable to those found in liquid motors thus permitting the rocket to accelerate slowly through the dense atmosphere and thereby utilize the available thrust to greater advantage in achieving a high peak altitude. Of such a design is the ARCAS rocket. Development of the ARCAS rocket was performed by the Atlantic Research Corporation under contract to the Office of Naval Research. This program which was initiated in 1958, was aimed at providing research scientists with an efficient low-cost vehicle for conducting experiments in the lower stratosphere at altitudes appreciably above those capable of being reached by balloons. The program goal was to obtain a small single stage rocket which could carry instruments similar to those used on balloons to a height of approximately 60 km where the instruments could then be expelled from the rocket and lowered by parachute, making measurements on the way down. Retarding the rate of fall of the instruments by parachute permits the use of sensors which have relatively slow response times and also allows the use of narrow-band telemetry by reducing the amount of data required to be transmitted per unit time. Tracking of the parachute also yields wind data except near the peak of the trajectory where the fall rate becomes excessively rapid. The ARCAS motor is a single-stage solid propellant motor using an end-burning cartridge-loaded grain cast from high-energy, aluminized, plastisol, rocket fuel. The basic dimensions of the ARCAS are 113 mm in diameter, 2.34 meter in length with a total weight of 35 kg including a five kg payload. When fired from sea level at a near vertical elevation angle carrying $3\frac{1}{2}$ kg of instruments attached to a parachute weighing $1\frac{1}{2}$ kg the ARCAS will reach an altitude of over 65 km. Above this altitude use of a parachute as a drag device to restrict the rate of fall is not practical due to the low density of the atmosphere. Fired from sea level without a parachute and only $3\frac{1}{2}$ kg payload the ARCAS has achieved an altitude in excess of 86 km. Approximately 5000 cm^3 of usable volume is available for parachute and instruments in the ARCAS.

As stated before, the end-burning feature results in a long burning time allowing a large portion of the total thrust to be delivered at high altitude where the drag is only a fraction of that found in the dense air sea level. Also by accelerating slowly through the dense atmosphere a minimum of aerodynamic heating is encountered thus making the heat shielding problem a minor one. One problem of the end-burning motor is that due to low thrust and slow acceleration the rocket must be given an initial boost in order to achieve stable flight upon leaving the launcher. One means of achieving this assist would be by use of a small booster rocket. This, however, degrades the simplicity of a true single-stage vehicle. In the ARCAS system this initial acceleration to stable flight is achieved in a distance of 15 ft without auxiliary boosters by employing a unique closed breach launcher which utilizes the initial exhaust gases to build up pressure behind a follower plate attached to the base of the rocket. The launcher then acts as a compressed gas gun and expels the rocket from the muzzle with a velocity in excess of 50 m/sec.

Use of the rockets described above has made possible a marked increase in the number of observations of meteorological parameters from the stratosphere. Cooperation among many interested agencies has even resulted in the establishment of a small synoptic network of rocket observations which may one day expand through international cooperation to become a world wide net work. Because of cost and ease of handling small rockets will probably be the only ones used extensively for synoptic observations, and they will probably be the most common rockets used for atmospheric research to the altitudes reached by them. Nevertheless, smallness in research rockets is not always desirable. There are times when the smallest motor that will carry the given payload weight to the desired altitude is not the best one to use. This may be true because of physical limitations on the minimum size of the instruments used as is often the case with optical equipment. Another consideration may be the power requirements of certain types of experiments that call for a high rate of information acquisition and transmission, and where it may not be economically sound to use very high capacity and expensive power sources just to permit use of a small rocket. Other considerations may also result in the decision to use a larger vehicle than indicated by a pure weight to altitude approach. Generally speaking, however, the rocket region is wide open to the small rocket user and this first step to the stars can be taken without elaborate launch complexes or astronomical budgets.

Multiple-Stage Solid-Propellant Sounding Rockets for Space Research

William H. Hansen and Frederick F. Fischbach

The University of Michigan Research Institute, Ann Arbor, Mich., USA

Aerodynamically stabilized sounding rockets with performance useful for space research can be constructed from existing solid-propellant motors. An appreciable portion of current high-altitude research is carried out with such vehicles. Three such systems, one capable of carrying 23 kg of payload to 160 km altitude, a second capable of sending 36 kg to 480 km, and a third which has sent 9 kg to 1770 km altitude, have been developed at The University of Michigan, and are described. All these systems have been characterized by low cost and high reliability. Design considerations, such as aerodynamic stability, interstage coupling, protection against aerodynamic heating, and launchers, are discussed. Also pointed out are future capabilities based on increasing availability of military solid-propellant rockets with considerably improved specific impulses and mass ratios.

Small solid-propellant rockets were first used for upper-air research by Van Allen [1] in 1952 with the launching of the Deacon rocket from a balloon in the well-known "Rockoon" combination. The advantages of simplicity, reliability, and low cost which may be achieved with solids were effectively demonstrated, and efforts to realize these advantages in an all-rocket system resulted shortly in the Nike-Deacon [2] and later in the Nike-Cajun [2, 3, 4], Terrapin [2], Nike-Asp [2], Exos [5], Strongarm [6], and others. In many cases the solid units had been developed for military purposes and were available at moderate costs due to high production. This is a key factor in the ultimate low cost of the sounding vehicle combinations described below. Fortunately, many of the units may be obtained commercially.

In 1955 the High Altitude Engineering Laboratory of The University of Michigan, with the valuable aid of the Pilotless Aircraft Research Division of the National Advisory Committee for Aeronautics, began its rocket vehicle work with the Nike-Deacon two-stage combination. The Deacon was shortly superseded by the Cajun. The Nike-Cajun was developed for the purpose of carrying the falling-sphere experiment [8] for upper-air density. The Nike-Cajun carries 23 kg to 160 km. Since the sphere instrumentation was simple, small, and inexpensive, a rocket vehicle having the same qualities would permit the combination to be used frequently and independently of the support of a major missile base—two features required for synoptic work. Another important factor in achieving the low-cost objectives of the Nike-Cajun was that the scientific research group itself could, and did, assemble and fire the rocket. In the continuing work with staged solid rockets, which has led to a 480-km vehicle and a 1900-km vehicle, the capability of firing by university research personnel has been largely maintained. Thus even in the case

Fig. 1. Nike-Cajun on launcher.

of the 1900-km Strongarm, which is physically not a small rocket, assembling and launching the rocket are largely within the capabilities of a university group.

During the IGY about half the rockets fired in the US program at Ft. Churchill, Canada, were Nike-Cajuns, and this type of medium-performance, two-stage rocket currently carries a significant portion of high-altitude payloads. The Nike-Cajun and Nike-Asp have been used to measure the pressure, density, temperature, composition, and winds to 100 km as well as to investigate cosmic rays, charge density, solar ultra-violet and X-rays, micrometeorites, auroral particles, magnetic field, and airglow.

The Nike-Cajun

The first stage of the Nike-Cajun (Fig. 1) is the booster from the Nike-Ajax anti-aircraft missile system and the second stage is the Thiokol Chemical Co. Cajun, a 15.2-cm-diameter, 244-cm-long motor which burns with a thrust of 3650 kg for 2.75 seconds. Both stages are fin-stabilized. The second stage simply sits on a cast coupling and separates by differential drag on burnout of the first stage. Ignition of both stages is accomplished on the launcher with an 18-second-delay squib allowing a suitable coast period following separation before the second stage begins to thrust. The simplicity of the Nike-Cajun is illustrated by the fact that in 1956 a group of five men from The University of Michigan took the vehicle and the falling-sphere experiment aboard the U.S.S. Rushmore (LSD-14) and fired five rockets

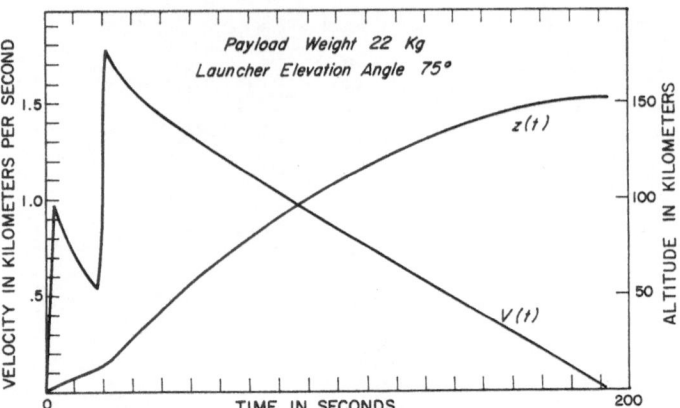

Fig. 2. Trajectory of Nike-Cajun launched at 75° elevation.

at various latitudes between 39°N and 65°N. The group did virtually all the work of assembly and launching as well as servicing the instrumentation and manning the data receivers.

Most of the more than 100 Nike-Cajuns flown have been successful. In a few instances where the compatibility of the instrumented nose cone to the vehicle received insufficient study, the flights were spectacular if not successful. Some payloads were long, heavy, and somewhat flexible. In others the nose cone had a 50% increase in diameter over the diameter of the second stage. The failures generally resulted from aerodynamic and/or structural divergence [9] with consequent break-up near the end of the first-stage burning. The light payloads (20 to 23 kg) reached altitudes of just over 160 km. The heaviest payload flown (54 kg) reached an altitude of 116 km. Fig. 2 is a plot of altitude and velocity *versus* time for a typical Nike-Cajun with a 23-kg payload launched at 75° elevation angle. Note that the peak velocity at second-stage burnout is about 1830 meters per second. The peak axial acceleration is about 70 times gravity. It was first thought that the problems of designing and constructing instrumentation that would operate under this acceleration would be very difficult. It has been demonstrated that, with thought and some ingenuity, these problems can be overcome. Transistorized circuits have helped greatly in solving the problems associated with high acceleration as well as the problems of making the instrumentation smaller. Most of the Nike-Cajuns have been launched from a modified Nike-Ajax system launcher. This is a rail-type launcher which provides about 3 meters of guidance, and has been used mainly because of its availability at numerous launching sites. The Nike-Cajun can be flown equally well from a zero-length launcher although the dispersion may be a bit larger. Only a modest amount of ground support equipment is required. A mobile crane, fork lift truck, or adjustable dolly is essential for loading the 544-kg first stage. The second stage with payload can be man-handled. The Nike-Cajun vehicle in general poses few problems from aerodynamic heating. With light payloads, say 23 kg, the fins (which are made from 14S-T4 aluminum alloy extrusion and are only 2.5 mm thick at the tip) are protected by adding a 0.78-mm Inconel— a nickel alloy with high melting temperature—leading edge cuff. This cuff is 5 cm wide and is riveted to the fin over a layer on Fiberglas cloth. An improved Cajun rocket called the Apache has recently been flight-tested. This rocket has virtually the same metal parts but with higher total impulse and longer burning time. The Apache has demonstrated in a single test flight its ability to take as a single stage a 16-kg payload to an altitude of 61 km. The Apache represents an increase in total

Fig. 3. Exos on launcher.

impulse of 24% over the Cajun. Development work is continuing toward greater impulse and longer burning times in the same general size and configuration. In conjunction with the Nike booster, the Apache is predicted to carry 27 kg to 275 km.

Exos

A second sounding vehicle developed by the Michigan group is Exos. This design resulted from a requirement to take an 18-kg payload to an altitude of from 320 to 480 km. A three-stage combination was selected, starting with the Honest John, a ground-to-ground artillery rocket yielding very high thrust for over four seconds. It is presently being produced in quantity and hence its cost is relatively low. The standard Honest John fins are used in the Exos. The second stage is again a Nike booster with cruciform fins having an area of 0.23 square meter per fin as in the Nike-Cajun. The third stage was originally a Thiokol Recruit rocket of high impulse and short duration. The instrumentation at time of Recruit burnout experienced an axial acceleration of nearly 175 times gravity. By the time the third Exos vehicle was fired, a new version of the Recruit, called the Yardbird, was substituted and the acceleration was reduced to approximately 80 times gravity. Fig. 3 shows

TABLE I. Flight Parameters of Exos
(37-kg nose)

Action	Time (sec)	Weight (kgf)	Drag (kgf)	Altitude (m)	Velocity (m/sec)	Acceleration (g's)
1st Ignition	0	2640	0	0	0	14.5
1st Burnout	4.4	1753	5443	1524	725	19
Separation	4.45	789	1762	—	—	— 3
2nd Ignition	29.4	789	66	13707	316	27
2nd Burnout	32.6	430	494	16401	1409	48
3rd Ignition	32.6+	207	166	—	—	41
3rd Burnout	35.9	86	195	23827	3369	100
Peak	393	86	0	645 km	0	— 1

the Exos on a rail type launcher. The vehicle takes off with 14.5 times gravitational acceleration and is suitable for zero length launching. The flight parameters are shown in Table I. The values given were calculated for a 90° launch angle. Table II gives the peak altitude reduction factor for various launch angles between 75° and 90°. The vehicle is fairly insensitive to changes in payload weight. Peak altitude loss for added weight is about 0.73 km per kg. Exos, with a burnout velocity in excess of 3000 meters per second at 19.2 km altitude, does present some aerodynamic heating problems. Fig. 4 shows calculated skin temperatures *versus* time at two places on the third stage. The upper curve shows temperatures of the 1.27-mm Inconel nose cone at station 15 just aft of the solid nickel nose tip, and the lower curve shows temperatures on the case magnesium flared skirt. In this instrumentation we used a heat sink approach to the nose-cone-heating problem. The tip itself was a 3.6-kg nickel casting separated from the nose cone by a porcelain insulator. The tip then was one side of a dipole telemeter antenna. The cone itself was a 1.4-mm Inconel sheet rolled, welded, and spun. A light-weight, bright-surface, aluminum reflective shield was placed around the instrumentation. The skirt serves the dual purpose of providing aerodynamic stability and acts as part of the coupling to the second stage. Fig. 5 shows the construction of the blast diaphragm type of coupling. The flared skirt on the forward stage and the coupling casting bolted to the aft stage are both threaded on the outside of the blast diaphragm. When the forward stage is ignited, the pressure of the exiting gas bows the diaphragm so that the threads become disengaged from the flared skirt, and a clean, rapid separation occurs. This system is generally used between stages which are fired in succession without a coast period. The large diameter of the skirt and the burned out preceding stage would cause large drag losses during a coast period. The flight tests in

TABLE II. Peak Altitude Reduction Factors

Launch Angle, °	Peak Altitude Reduction Factor
90	1.000
87.5	0.995
85	0.981
82.5	0.957
80	0.918
77.5	0.883
75	0.834

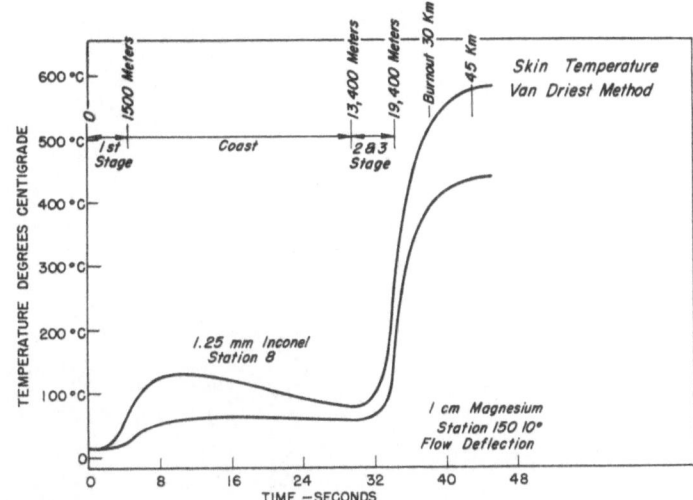

Fig. 4. Skin temperature as a function of time.

two of three cases have demonstrated that the vehicle more than meets the design specifications. The first shot launched at 75° elevation angle carried a nose-cone weight of 36 kg to 386 km, while the second carried the same weight to 476 km. The latter was launched at 80° and exceeded the prediction by about 13 km. The third test was a failure due to some malfunction during third-stage burning. It is believed that either the nozzle came off or that the case burned through at the aft end. The second-stage hardware has proved to be very reliable through more than 100 firings as Recruits. The longer-burning Yardbird propellant caused nozzle failure in static firings. Corrective measures were taken and successful static firings and a single flight test were accomplished prior to the failure.

The Exos perhaps does not fall into the category of a small rocket; yet it can be handled, assembled, and flown by a crew of modest size. The launcher shown in Fig. 3 is one of the first experimental Honest John launchers. Originally elevation of the loaded launch rail was accomplished with a crane. When the desired ele-

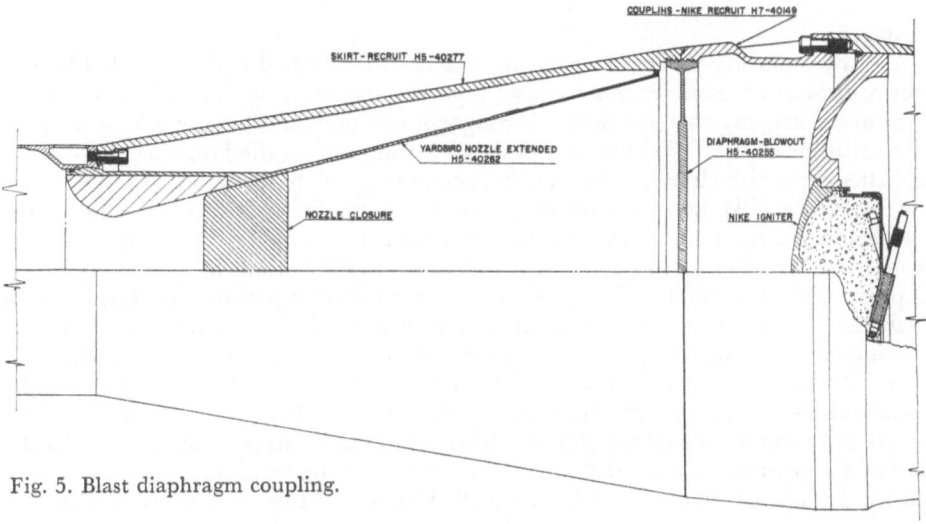

Fig. 5. Blast diaphragm coupling.

Fig. 6. Strongarm on launcher.

vation angle was reached, the A-frame strut was pinned to rails attached to the concrete launch pad. The launcher was modified by shortening the launch rail to a guided travel length of 3.6 meters and adding a motor-driven pump and hydraulic rams to elevate the launching rail. The launcher is not trainable in azimuth, which is a disadvantage at some firing ranges.

Strongarm

A third sounding rocket, Strongarm, has been developed and assembled by our group. Insofar as possible, this too uses high-production, military rocket engines as a means of keeping the cost down. Strongarm was designed to take a 9-kg payload to an altitude of over 1600 km. Fig. 6 shows Strongarm on the launcher. Like Exos, its first stage is the Honest John and its second stage is the Nike booster. The third stage is also a Nike booster with 0.185 square meter fins. The fourth stage is the Yardbird and the fifth is Jet Propulsion Laboratories' Scale Sergeant. The last stage is the same engine, with minor hardware modification, as that used in the Explorer series of satellites. Table III shows selected flight parameters. The burnout of the final stage occurs at 35 seconds at an altitude of 53 km and velocity of approximately 5300 meters per second. Because we depend entirely on aerodynamic forces for stability the powered portion of the flight must take place at altitudes where sufficient dynamic pressure can be obtained. With velocities well over 3000 meters per second, this means that the heat-input rates will be considerable. In the case of Strongarm, not only did the nose cone have to be considered and protected, but the fifth-stage motor as well. The wall thickness of this motor is only 0.58 mm

TABLE III. *Flight Parameters of Strongarm*
(9-kg nose)

Action	Time* (sec)	Altitude (m)	Velocity (m/sec)	Weight (kgf)	Mach No.	Dynamic Pressure (kgf/m²)	Reynolds No. per meter (millions/m)	Drag (kgf)	Acceleration (g's)
1st Ignition	0	20	0	3220	0	0	0	0	12
1st Burnout	4.375	1260	588	2291	1.75	18776	35.9	3220	16
Separation	4.425	—	—	1398	—	—	—	2578	—3
2nd Ignition	6.375	2368	535	1398	1.62	13787	29.5	1534	14
2nd Burnout	9.375	4670	1002	1038	3.12	37582	45.0	2960	18
Separation	9.425	—	—	805	—	—	—	2957	—5
3rd Ignition	23.375	15700	656	805	2.22	3769	7.9	257	28
3rd Burnout	26.375	19177	1721	446	5.83	14993	11.8	735	50
4th Ignition	26.375+	—	—	212	—	—	—	259	40
4th Burnout	29.625	27450	3559	90.8	11.88	17360	6.6	250	90
5th Ignition	29.625+	—	—	40.4	—	—	—	81	18
5th Burnout	35.875	54494	5325	17.7	15.92	932	0.16	4.5	—1.3
Peak	818	1958000	0	17.7	0	0	0	0	—0.6

* Time is taken to nearest 1/8 sec for computer solution.

and the propellant grain is cast in after the inner surface has been coated with a bonding thermoplastic liner. Because the case would be subjected to extensive heating for 30 seconds prior to ignition of the motor, it was feared that the liner would become liquid and the grain would be loose, thus causing burning over too much of the grain surface with subsequent over-pressure and rupture. The maximum allowable temperature prior to ignition was 120 °C. The problem was solved by using an ablative coating. Sheet Teflon (E.I. Dupont's Tetrafluoro Ethylene) 1.27 mm thick was glued to the entire exposed surface. The Teflon ablates at approximately 600 °C and provides heat protection in three ways. First, Teflon is a moderately good insulator and delays the heat transfer by conduction. Secondly, considerable heat is dissipated in the ablation process, thirdly, the gas product of ablation increases the boundary-layer thickness.

The instrumentation flown in Strongarm for the measurement of ion density requires two fairly large loop antennas. The antennas were mounted within the nose-cone shell so the nose cone had to be made of a material which would be transparent to the radiofrequency transmission. Molded Fiberglas has this property and is structurally more than adequate. However, Fiberglas will char if subjected to sufficient heat and the charred surface is opaque to the frequencies used. Teflon sheet 1.78 mm thick was applied over the Fiberglas to prevent charring. The effectiveness of this approach was demonstrated in the first test when signals were received throughout the entire 25-minute flight. Fig. 7 shows the calculated and measured temperature *versus* time at station 28 on the nose cone. The second- and third-stage fins had Inconel leading-edge cuffs for heat protection and the fifth-stage flared skirt was cast from a rare-earth magnesium alloy which exhibits better high-temperature properties than the common alloys.

Strongarm requires the same launching and handling equipment as Exos. All its flight tests to date have been made from the versatile zero-length boom-type launcher shown in Fig. 6. In fact, this single launcher is easily adapted to all vehicles discussed in this paper.

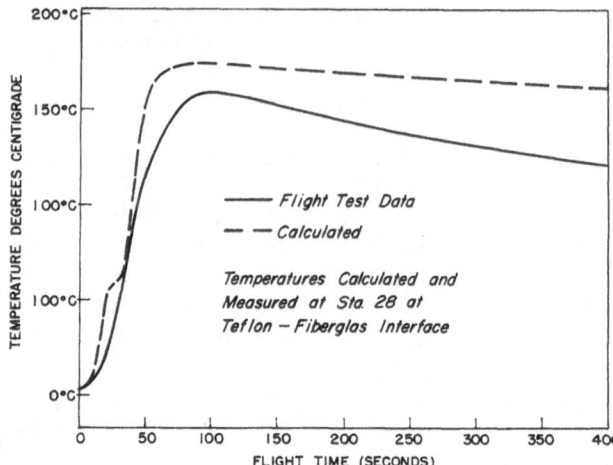

Fig. 7. Calculated and measured
temperature *versus* time.

A fourth vehicle is now being designed which will take payloads of 34 to 57 kg to the 220- to 320-km altitude range. It is a two-stage vehicle called the Nike-Yardbird and is expected to have all the simplicity and ease of handling of the Nike-Cajun system. The first test flight will be made this fall.

Conclusion

The University of Michigan program of staging solid-propellant units in sounding rocket combinations for use in our upper-atmosphere program has been an interesting and useful exercise. We hope that our experience will encourage university groups in other countries to undertake high-altitude research.

The considerable aid given us by the Pilotless Aircraft Research Division of the National Advisory Committee for Aeronautics is acknowledged. We are also indebted to the US Air Force Combridge Research Center for financial support and cooperation in the development of Nike-Deacon, Nike-Cajun, and Exos. Financial Support for Strongarm and Nike-Yardbird development was supplied by Ballistic Research Laboratories of the US Army Ordnance Corps.

References

[1] *Boyd, R. L. F.,* and *Seaton, M. J.,* Rocket Exploration of the Upper Atmosphere, Pergamon Press, Ltd., London (1954).
[2] *Newell, H. E., Jr.,* Sounding Rockets, McGraw-Hill Book Co., Inc., New York (1959).
[3] Annals of the IGY, Vol. VI, Part II, p. 82, Pergamon Press, Ltd., London.
[4] *Hansen, W. H.,* and *Fischbach, F. F.,* The Nike-Cajun Sounding Rocket, The University of Michigan Research Institute Report 2453-1-F, Ann Arbor, March, 1957.
[5] *Hansen, W. H.,* and *Fischbach, F. F.,* The Exos Sounding Rocket, The University of Michigan Research Institute Report 2595-1-F, Ann Arbor, December, 1958.
[6] *Hansen, W. H.,* and *Fischbach, F. F.,* The Strongarm Sounding Rocket, The University of Michigan Research Institute Report 2816:004-1-F, Ann Arbor, May, 1960.
[7] I.G.Y. Rocket Report Series, No. 3, National Academy of Sciences, Washington, 30 September 1959.
[8] *Jones, L. M., et al.,* Upper-Air Density and Temperature: Some Variation and an Abrupt Warming in the Mesosphere, Journal of Geophysical Research, *64,* 2331—2340 (1959).
[9] *Arbic, R. G., White, G.,* and *Gillespie, W., Jr.,* Some Approximate Methods for Estimating the Effects of Aeroelastic Bending of Rocket-Propelled Model Booster Combinations, NACA RM L53A08, Washington, March 27, 1953.

Small Sounding Rockets under Development by the Indian Astronautical Society

B. Ramesh Rau

Indian Astronautical Society, Mysore, India

The paper reviews the details of the various rockets that are under investigation. The first part of the paper shows the advantages of using small sounding rockets with limited payloads over bigger units with the same payload capacity. A comparative study of the large and small units is tabulated to show the efficiency of the smaller ones. The study can be considered as important because a country like India cannot afford large liquid or solid propellent rockets, particularly as the vehicles considered here are very economical. The study also indicates the type of performance that can be obtained even with small rockets employing fuels of moderate power. Tabulated data on the performance of staged vehicles of the same calibre are also given.

The rockets of today cover a wide range of sizes, from small signal rockets to the large missiles. Although the emphasis is now on large units, there will always be a requirement for smaller units as well. In the constant endeavour to design better rockets and discover new fuels, the possibility of using small solid propellent rockets for upper atmosphere research has almost been passed over. As the cost of large liquid and solid propellent rockets run into astronomical figures, we are forced to turn our attention to cheaper and consequently smaller units. The simplicity of small solid propellent units is one of their main advantages. They can be constructed and tested by far more easily than the large motors of today. They could very well use the often tested and reliable propellents. Of course, the more powerful propellents under trial could be used and would be very advantageous in the long run, though there will be the usual expenditure of time and money on research.

The small units that are to be considered in this paper are designed to use a composite plastic propellent like the GALCIT-58 propellent developed by the Jet Propulsion Laboratory. It is simple and does not need a large processing plant and other complicated equipment for its manufacture. An exhaust velocity varying between 5600 to 6000 ft per second could be obtained with chamber pressures up to 1800 psi. Though rocket units of diameters up to 12 in. have been considered here, these can be classified as small units on the basis of their short lengths, viz. three feet. The British SKYLARK, which can be considered as a large rocket weighs around 2500 lb at take-off and is capable of 100 miles altitude. It has a payload capacity of 150 lb. This performance cannot be equalled with small units, but they can take smaller payloads to 100 miles or more. They are lighter, easier to handle and do not require large launchers. The same ratio of the payload of the motor weight can also be maintained. This ratio for the SKYLARK is only 0.06, while that for the smaller units are higher, viz. about 0.1. Even a comparison between a rocket motor with a length of 6 ft and diameter of 12 in. and another motor of the same diameter but

TABLE I. Performance Comparison of Single-Stage Rocket Units

Length of motor (ft)	Diameter of motor (in.)	Take-off weight (lb)	Payload (lb)	Altitude (miles)
3	6	125	40	20
3	6	105	20	50
3	6	95	10	80
3	8	183	40	60
3	8	163	20	100
3	8	153	10	130
3	12	386	96	75
3	12	346	46	130
6	12	700	94	100

only 3 ft in length shows clearly that the smaller units will turn in a better performance. The larger unit weighing 700 lb and with a 96 lb payload will reach 100 miles, or 130 miles with a 46 lb payload. This is due to the fact that the high mass-ratio and the consequently high velocity also results in increase of air drag. The small unit weighing only 350 lb will with a 46 lb payload reach an altitude of 130 miles. Even with a 96 lb payload it will obtain data up to 80 miles altitude. The increased air drag due to the shorter burning time is, however, overshadowed by the reduction of the loss of velocity due to gravity for the same reason. But there is also a limit to decreasing the length of the motor, as at some stage the increase in air drag overtakes the gain in velocity. Here is an example of this effect: A rocket motor of length 1.5 ft and 12 in. diameter, will reach an altitude of 100 miles only even with a reduced payload of 20 lb while a 3 ft length motor of the same diameter will

TABLE II. Comparative Performances of Two-Staged Vehicles

Details	First Stage		Second Stage
Length	3 ft		0.75 ft
Diameter	12 in.		12 in.
Take off weight	400 lb		100 lb
Payload	100 lb		20 lb
Altitude		360 miles	
Length	3 ft		1 ft
Diameter	8 in.		8 in.
Take off weight	500 lb		70 lb
No. of motors	3		1
Payload	70 lb		20 lb
Altitude		340 miles	
Length	3 ft		1 ft
Diameter	12 in.		8 in.
Take off weight	370 lb		70 lb
Payload	70 lb		20 lb
Altitude		300 miles	

reach 130 miles with 46 lb payload. The larger the diameter, the more complex are the problems that confront us. But when the burning period is reduced to a certain level, some of the problems are fairly easily solved. As motors with diameters of 8 to 9 ft have-been tested successfully elsewhere, it should not be difficult to test motors of the size discussed here.

Smaller diameter motors can be reliably employed to push up smaller payloads up to 100 miles. With these small payloads the larger units are capable of altitudes of 200 miles or more. These smaller units of diameters 6 to 8 in. with payload capacities varying between 10 and 40 lb can reach altitudes between 20 and 130 miles. A few examples of this type have been designed and their expected performances are listed in Table I. Here again only lengths of 3 ft are considered as the propellent used is slow-burning. The longer units of this type tend to curtail the velocity rather than increase it. This is mainly due to increased burning time and the resulting increased gravitational loss.

The 6 in. rocket could be constructed at a cost of less than Rs. 400/— (£ 30/—) and the larger units (12 in.) would cost less than Rs. 2000/— (£ 150/—). The two-stage vehicles of the sizes mentioned here would cost about Rs. 2500/— (£ 185/—).

Stages vehicles would work out quite satisfactorily and would make handling quite simple if they are fairly small. All the three vehicles designed have a payload capacity of 20 lb. The heaviest of them weighs only 500 lb. If the payload is increased to 40 lb their altitude ranges will be between 200 and 250 miles.

Appendix

The experimental investigation by the Indian Astronautical Society has been so far on small scale powder rockets. More than a dozen rockets were, during 1958—59, fired with comparative success. These small rockets were launched to various altitudes between 1000 ft to 5000 ft. A two-step rocket went up to an estimated altitude of 60 000 ft. It carried a 6 lb payload which included a simple device to ignite a magnesium charge automatically when the maximum altitude was reached. Another large rocket, 13 ft long and 4 in. in diameter, fell back after reaching an altitude of only 5000 ft. This low performance was due to a leak in the combustion chamber.

Later the principle of using shaped charges to effect a controlled angled flight path for solid rockets were studied. Three different configurations for the charges were designed and tested. The tests indicated that this type of system was a simple method of directing rockets along a pre-determined flight path. The method of separating the expansion section of the nozzle of the motor and thus deflecting the thrust to obtain the necessary inclination in flight was also tried. Owing to the complexity of this system over the former, the investigation was discontinued.

Lately, testing of composite case bounded plastic propellents have been undertaken. A few static tests have been conducted. The next step in this connection is the construction and testing of rockets using this fuel. An 8 in. diameter and 1.5 ft long rocket (length without instrument section) is under construction and will be tested shortly. It is equipped with a transmitter and power supply units. It may also contain a sodium evaporator to release a sodium cloud during its free flight. With a 5 lb payload this rocket has an altitude capability of 100—150 miles. This will be the first large scale experiment to be conducted in 1960. Depending on the performance of this rocket further rockets will be designed and tested. Some of the rockets previously mentioned are the ones that are under study for development and launch. Methods of launching instrument capsules to great altitudes by other than rocket propulsion are under active consideration, and any fruitful results that may be obtained will force us to cancel all scheduled rocket experiments.

Aurora — A Swedish High Altitude Rocket Project Study

Lars Henrik Ågren, Royal Swedish Air Force Board, Stockholm, Sweden

Olle Ljungström, SAAB Technical Office, Stockholm, Sweden

Åke Hjertstrand, Research Institute of National Defence, Dept. 1, Sundbyberg, Sweden

Project Aurora is a study regarding possible Swedish built sounding research rockets utilizing existing rocket motors originally developed for other purposes. The main elements of the Aurora rockets are one bigger and one smaller long-time burning rocket step and short-time solid propellant booster rockets. By combining, three variants are obtained, namely

Type	Launching weight (kg)	Payload (kg)	Corresp. summit altitude (km)
HR-1	845	5—25	400—310
HR-2	755—845	25—115	185—115
HR-3	190	5—15	80—65

After minor modification the HR-3 should be able to reach about 100 km height. The rockets are equipped with parachutes for recovery of the bigger and expensive liquid propelled stage one of the HR-1 and HR-2 as well as the separated instrumented payloads. In certain cases increasing the payload weight by omitting the parachute could make non-rescuing the payload worthwhile. The rockets are unguided and are aerodynamically fin-stabilized and also spin-stabilized with relatively low spin velocities between 0.7—2 rps. Maximum acceleration values are modest for HR-1 and HR-2, about 7 g, and for HR-3 about 17 g. The Aurora rockets could be launched from fixed or mobile ramps. Depending on which alternative is chosen the time space from a possible order to delivery of the first serial rockets should take 1½—2½ years. Prototype launchings should be possible already after 1—1½ year. Following the combination system on which Project Aurora is based, it is possible to start with one variant giving heights up to 60—100 km and later extending the program up to about 400 km height.

Some years ago the question was raised in the project study group of the Swedish Interplanetary Society to devote part of its efforts to an investigation concerning the optimization of high altitude sounding rocket projects. At that time, however, there seemed to be very little interest for such a study, partly owing to the extensive research already conducted elsewhere, and partly because of the high costs involved in starting more or less from scratch.

The International Geophysical Year aroused increased interest also for high altitude and space research. It seemed very likely that even modest enterprises would gain in importance and be carried on for still many years and in all possible areas. With this in view and as there are among the members of the Society a

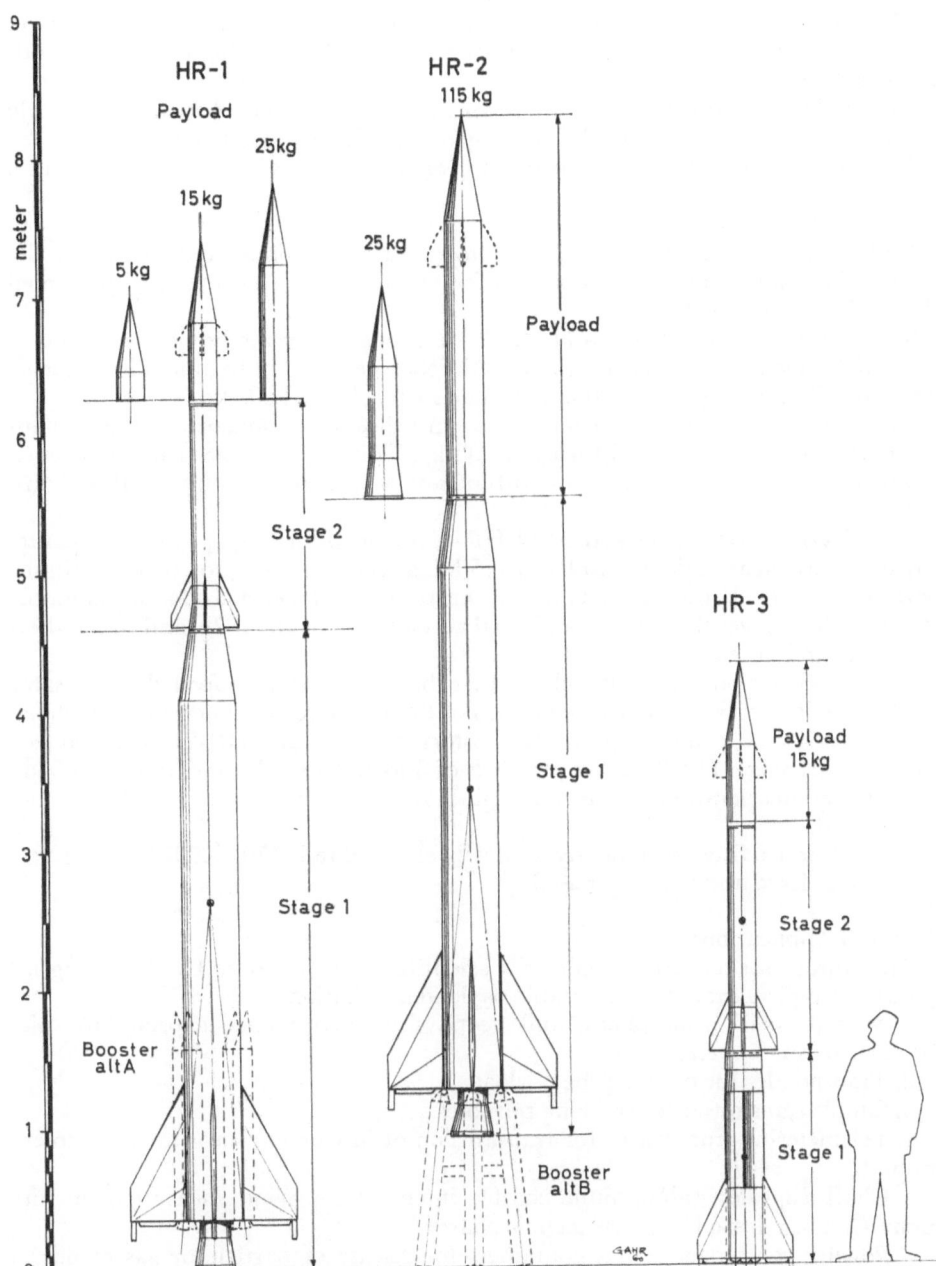

Fig. 1. Aurora rocket vehicle and payload variants.

number of experts in the fields of propulsion, aerodynamics, vehicles *etc.*, it was suggested in 1958 to undertake a study of a sounding rocket vehicle mainly depending on already existing Swedish rocket motors.

Although it was entirely clear from the beginning that this would exclude a general theoretical study of performance optimization, it soon became evident that this hardware approach indicated some promising possibilities. In the beginning

37

of 1960 the group had the pleasure to present before the Swedish Committee on Space Research a preliminary study of a proposed all-Swedish high altitude sounding rocket system.

This rocket system consists of a number of interchangeable elements to provide for more or less ambitious height and payload performances. The project was for the obvious reason of Sweden's northern geographical position christened Aurora.

General System Description

Project Aurora is based on rocket motors and other components originally developed for military purposes. Three different rocket have been projected, named HR-1, HR-2, and HR-3 (Fig. 1).

The two larger types HR-1 and HR-2 are propelled by the liquid rocket motor VR 3 developed by Svenska Flygmotor Aktiebolaget in Trollhättan. The sea-level thrust of VR 3 is 2210 kp, at 20 km altitude increasing to 2600 kp.

The HR-1 version has a second stage, propelled by a smaller solid-propellant rocket motor, with a sea-level thrust of 235 kp. This second stage is able to carry a payload of 5 kg up to over 400 km altitude and a payload of 25 kg to 310 km altitude.

The HR-2 is a one-stage variant of HR-1, intended for large payloads. Characteristical performance data: 115 kg to 115 km altitude or 25 kg to 185 km altitude. Both types have 4 booster rockets with a thrust of 1150 kp each and a burning time of 2.2 sec. This gives the vehicles an initial acceleration of about $7g$, and thus a short launcher can be used.

The HR-3 version is considerably smaller than the former, propelled by the same solid-propellant rocket motor as stage two of the HR-1 version and has three short-time burning rocket motors (same as boosters for HR-1 and HR-2) arranged as a separate first stage. HR-3 carries payloads of 5 to 15 kg to altitudes of 80 to 65 km respectively. Acceleration at launch is high ($17g$) and a simple zero-length launcher can be used.

All rockets are fired from nearly vertically elevated rails (70°—90°). The system can be both fixed and movably based.

Research Applications

The Aurora rockets are intended for scientific investigations. The following list gives some typical examples of suitable payload installations:

1. Miscellaneous atmospheric and geophysical instruments, capsules for biological experiments, etc..

2. Transponder for tracking the payload.

3. Simple transmitter for tracking purposes.

4. Telemetering transmitter for transmission of instrument measurements to the ground.

5. Chaff capsules (aluminium chaff), intended for wind measurement with radar. Can be released by a pyrotechnic charge.

6. Powder containers or gas containers for release of particle or gas clouds at high altitude.

7. Flash lights (magnesium), intended for geodetic measurements.

8. Aerodynamic test models (e. g. sphere or blunted cone), intended for hypersonic flight research and re-entry experiments, for this purpose equipped with an extra solid propellant rocket motor for acceleration downwards, perhaps also extendable stabilizing fins, and incorporating a transponder or telemetering unit for tracking and data transmission respectively.

A section of the atmosphere in Fig. 2 illustrates some applications.

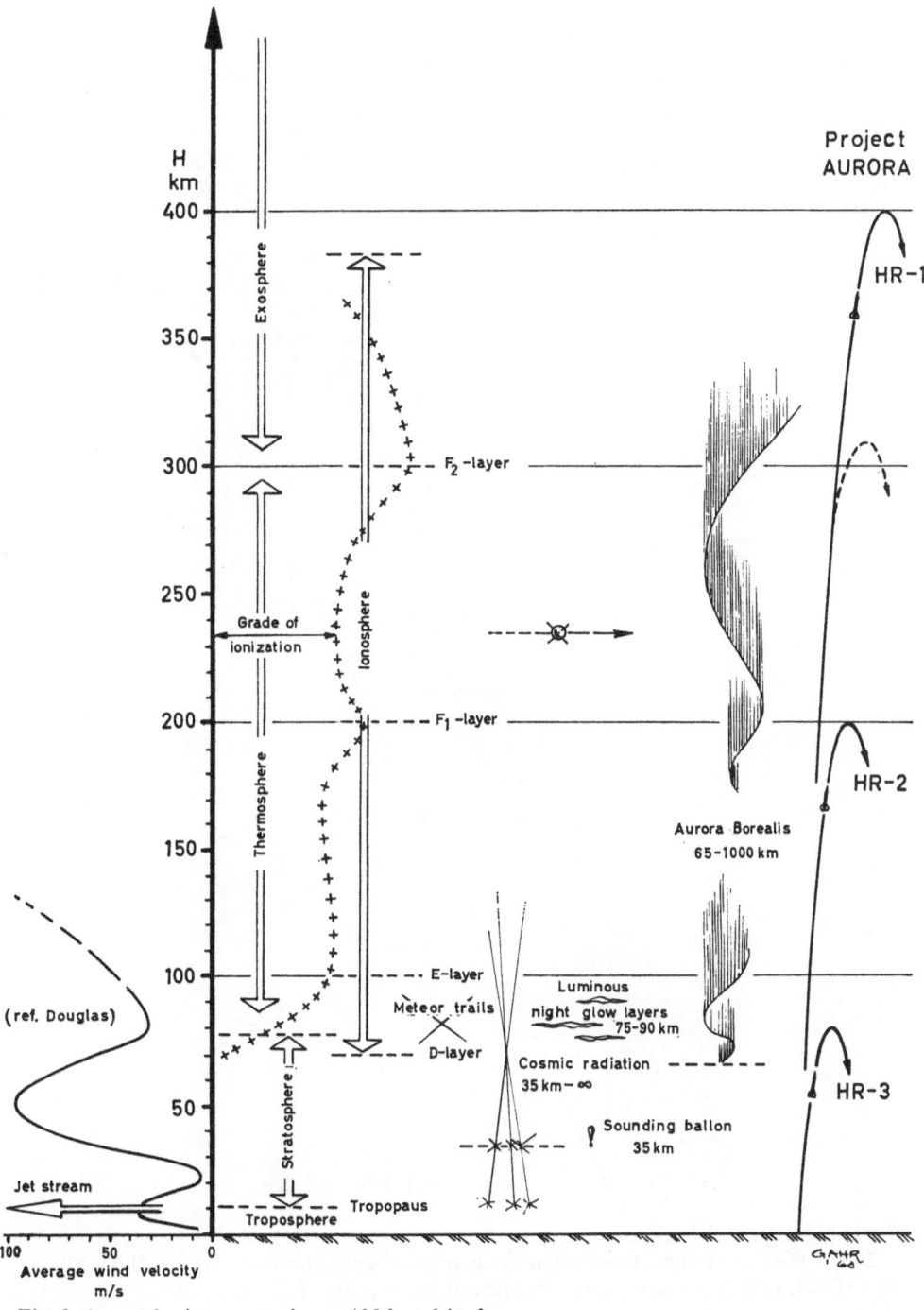

Fig. 2. Atmospheric cross section to 400 km altitude.

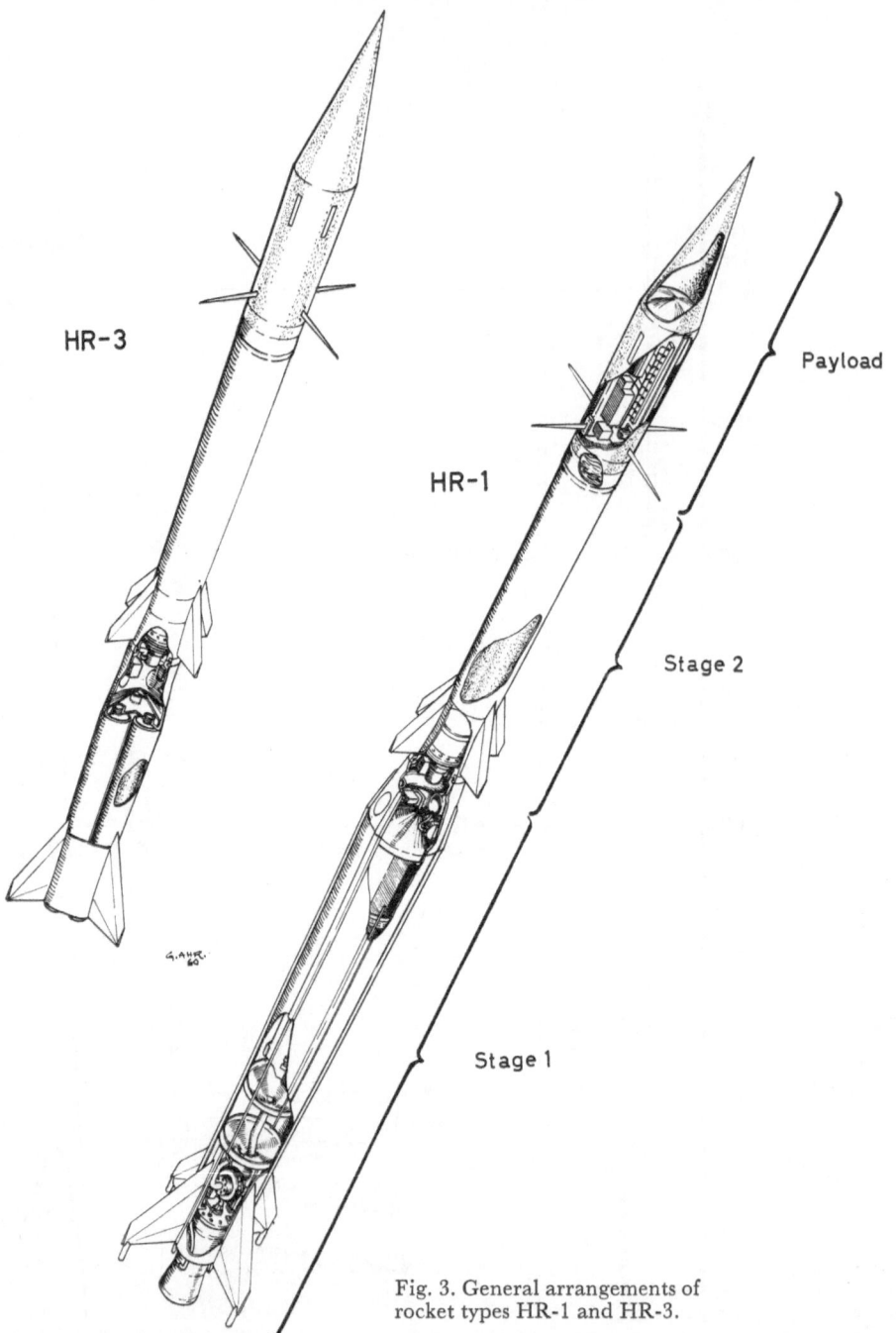

Fig. 3. General arrangements of rocket types HR-1 and HR-3.

Rocket Design and Performance

The following design description deals mainly with the two-stage rockets HR-1 and HR-3, which contain all the important basic units of the Aurora system.

Due to the preliminary character of the design study, detailed analysis of the kinetic heating conditions in the structure as well as dynamic load analysis (launch and high altitude wind shears, *etc.*) have not been completed. However, the structural weight estimate is quite conservative in order to cover all reasonable dynamic load cases, kinetic heating effects, *etc.*.

TYPE HR-1, THE LARGE TWO-STAGE ROCKET

Stage one. This stage contains the following main parts: The liquid rocket motor installation, propellant tanks integrally arranged in the rocket main body, stabilizing fins, and at the top a compartment for housing equipment and the second stage separation unit. (Figs. 1, 3.)

The liquid rocket motor utilizes the propellant combination high test hydrogen peroxide (HTP) as oxidizer and kerosene as fuel. It was originally designed as an aircraft powerplant with a high degree of thrust variability. The propellants are fed into the combustion chamber by means of a turbo-pump, driven by decomposed HTP from the high pressure side of the oxidizer pump. The main propellants are fed to the pump unit partly by gravity, partly by means of pressurization of the propellant tanks with nitrogen gas from a toroid-shaped high pressure container mounted in the conical compartment at the top of stage one. The tanks are pressurized to 2 atm which not only serves for feeding the propellants but also has a favourable influence on the strength of the integral tank structure itself.

The propellant flow rate is approx. 11 kg/sec during the burning period (41 sec) and includes the HTP supply to the pump turbine.

Compared with the somewhat more complex aircraft rocket engine the Aurora version has been stripped down to a lower weight (65 kg complete with turbopump unit and miscellaneous equipment, compared with 75 kg for the aircraft version).

The rocket motor is installed at the base of the rocket and is mounted with four fittings, bolted to the top cover of the combustion chamber and attached to four longerons in the rocket's shell arranged in line with the root fittings for the four stabilizing fins. Directly on top of the rocket motor installations is the smaller kerosene tank with conical bulkheads at the top and bottom. On top of this tank is the considerably longer HTP-tank. Both tanks are arranged integrally in the rocket shell structure. The tank cylinder is made of welded 1.5 mm aluminium alloy, stiffened externally by eight longitudinal hat stiffeners (adhesive bonded to the skin) and internally by four radially mounted vertical partition bulkheads, serving for prevention of propellant sloshing. The external hat stiffeners can be used as protective covers for electrical cables running past the propellant tanks. The HTP supply to the rocket motor passes through a central tube in the kerosene tank.

The conical structure section at the top of stage one houses the bulk of the control system of stage one, and is built as a ring-stiffened conical magnesium shell, 3 mm thick. The upper bulkhead of the HTP-tank is shaped as a pointed cone, which apart from having the favourable effect of reducing thermal stresses built up between the shell cylinder and the bulkhead also makes possible a clean escape for the exhaust gases from stage two when ignited. At the first instance of separation, these gases escape through four portholes in the cone shell structure, opened up by the action of the internal pressure built up.

The second rocket stage is attached to the first stage by means of a circumferential tie band, holding together circular flanges in each unit. Separation is initiated by release of this tie band with one or two cartridge bolts.

Stage two. This stage consists mainly of a slightly modified solid propellant rocket motor utilizing an end-burning propellant. The rocket's steel casing has an internal insulating liner, making external insulation against aerodynamic heating superfluous. The exhaust nozzle has been modified with increased nozzle exit area (increased expansion ratio) for better performance at operating altitudes. Ignition is arranged through a time relay, which is put into operation when stage one is started. The upper bulkhead of the solid propellant motor casing has been modified for attachment of the payload, cone-cylinders of varying length and weighing from 5 to 25 kg.

Boost rockets. In order to achieve a sufficiently high velocity at release from the launcher without making it too long and difficult to handle, a booster arrangement has been found to be necessary.

Among available solid propellant rocket motors, one of the most suitable for this purpose is the rocket motor of the 13.5 cm Bofors rocket projectile. With four rocket motors of this type, the boost thrust increment will be 4600 kp, making the total thrust at launch $F_{tot} = 4600 + 2210 = 6810$ kp, whereby the net acceleration at launch becomes 7 g (as compared with 1.9 g without boosters). Two different booster mountings have been studied as shown in Fig. 1. The best arrangement incorporates two pairs of rockets mounted directly on the outside of the rocket body, diametrically opposite each other, between the rear stabilizing fins. The four nozzles are mounted with slightly inclined thrust lines, pointing towards the c. g. of the complete rocket system at launch. The booster rockets are secured in their thrust transfer fittings with shear pins, which are arranged to be sheared by mass forces so that the rockets will be free to slide off rearwards after burnout at approx. 170 m altitude.

Nose section with payload. The nose section, mounted on top of the second stage, can be arranged to separate from this stage at a suitable moment during ascent or descent. The shell structure is made of 2.5 mm heat resistant reinforced plastics, preferably phenolic-glass fibre laminate, in order to protect the instrumentation against the kinetic heating effects. The internal frame structure is made of light alloys.

Recovery of stage one. A parachute can be stowed in the rear of the main rocket body in the cylindrical space between the rocket's cylinder wall and the rocket motor. The parachute system consists of a small first stage parachute, which is released barometrically and pulls out a reefed main parachute of approximately the same size as a personnel parachute. The empty weight of stage one is approximately 205 kg including the 15 kg parachute installation, and it is carried down to earth with the tank structure pointing downwards to collapse upon impact, leaving the VR 3 rocket engine installation intact.

TYPE HR-3, THE SMALLER TWO STAGE ROCKET

Both stages have solid propellant rocket motors. Stage one consists of three clustered 13.5 cm Bofors rocket motors of the same type as the boosters used with type HR-1, fitted with three stabilizing fins at the base. The three nozzles are slightly inclined so that a spinning torque is simultaneously obtained (Figs. 1, 3).

Stage two is identical with stage two in type HR-1, and is attached to stage one in the same way with a circumferential tie band.

SEPARATION AND RECOVERY OF STAGE TWO

For all three Aurora rocket models, the same separation system design and geometry is used between stages one and two (types HR-1 and HR-3) and between stage one and the payload (type HR-2), which makes possible a flexible choice of various combinations.

The separation between stage one and two in type HR-1 is initiated by a thrust sensing unit which closes the stage two rocket igniter circuit when the thrust of stage one falls down at burnout, and at the same time the separation mechanism

TABLE I. *Principal characteristics of the Aurora sounding rockets*

	HR-1	HR-2	HR-3
Dimensions (m)			
Length, max total	8.35	8.3	4.3
1st stage	4.25	4.25	1.5
2nd stage (incl. payload)	2.8	—	2.8
Diameter, 1st stage	0.4	0.4	0.26
2nd stage	0.20	—	0.20
Fin span, 1st stage (tip to tip)	1.2	1.2	0.8
2nd stage	0.48	—	0.48
Weights (kg)			
Launching weight	847	846	192
Payload	(5) 15 (25)	(25) 115	(5) 15
Total weight of 1st stage	637	637	76
Empty weight of 1st stage	197	197	
Propellant of 1st stage	440	440	
Total weight of 2nd stage	101	—	101
Booster	94	94	—
Performance			
Thrust, 1st stage (H = 0), kp	2210	2210	3450
Duration of thrust, sec	41	41	2.2
Thrust of booster, kp	4600	4600	—
Duration of thrust, sec	2.2	2.2	—
Thrust, 2nd stage (H = 0), kp	235	—	235
Duration of thrust, sec	40	—	40
Initial acceleration, g	7.0	7.0	17.0
Maximum acceleration, g	7.8	7.8	20.0
Total mass-ratio (incl. payload)	7.30	2.71	1.66
Burnout velocity, 1st stage, m/s	1350	1345	405
Burnout altitude, km	23.5	25.5	0.435
Burnout velocity, 2nd stage, m/s	2250	—	950
Burnout altitude, km	98	—	24.5
Peak altitude, km	354	115	68
Peak time, sec after launch	229	174	141
Guidance and control features			
Pitch and yaw	Spin-stab.	Spin-stab.	Spin-stab.
Roll	None	None	None

is released by pyrotechnic breaking bolts. If separation of the payload from stage two of rockets type HR-1 and HR-3 is required, this can be arranged with a barometrically or time-released breaking joint.

Stage two with payload or the payload alone can be recovered by means of a small parachute, stowed in the payload compartment, *e. g.* in the nose cone. The empty weight of stage two is 65 kg including 25 kg payload. A suitable recovery parachute for this unit complete with release mechanism weighs approx. 3.5 kg. For the 25 kg payload alone, this can be reduced to approx. 2 kg.

PERFORMANCE

The principal characteristics of the different Aurora rocket types are listed in Table I. From dimensions given herein and the configuration shown in Fig. 1 the drag of the different rockets has been estimated for both the powered and the coasting phases.

The predicted performance of the different rocket versions is given in Table I and Figs. 4 to 8. Fig. 4 shows the variation of acceleration, velocity, and altitude

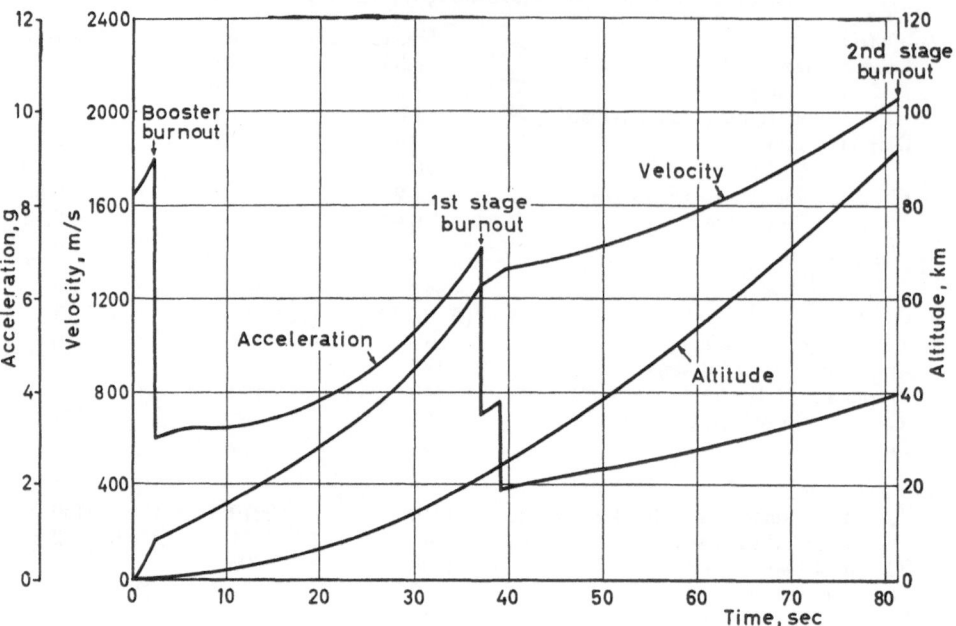

Fig. 4. Large two stage rocket HR-1 with 25 kg payload. Estimated acceleration, velocity, and altitude *versus* time.

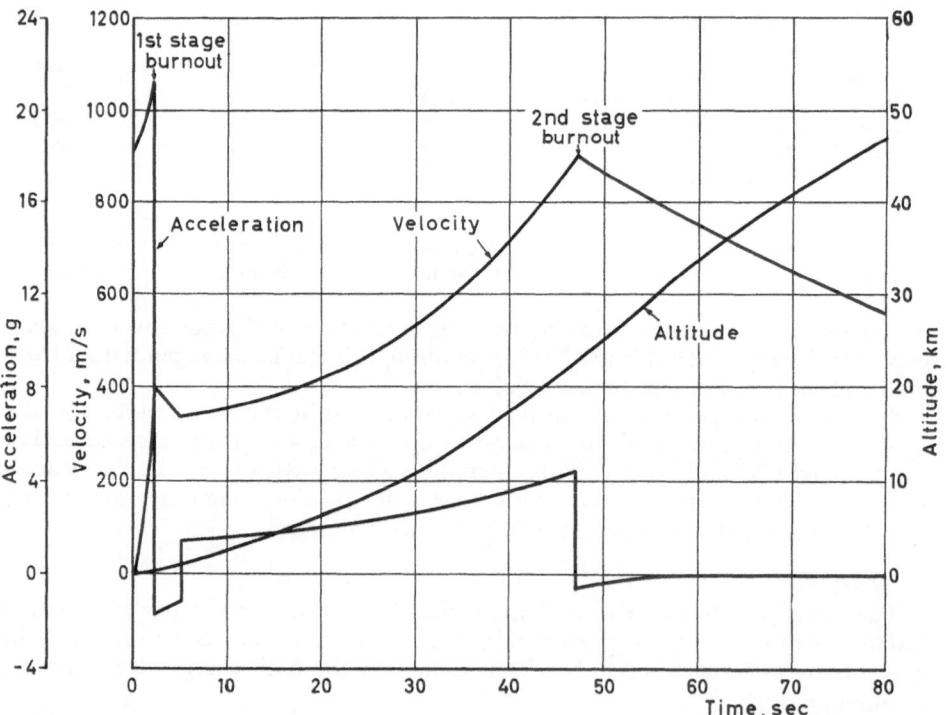

Fig. 5. Small two stage rocket HR-1 with 15 kg payload. Estimated acceleration, velocity, and altitude *versus* time.

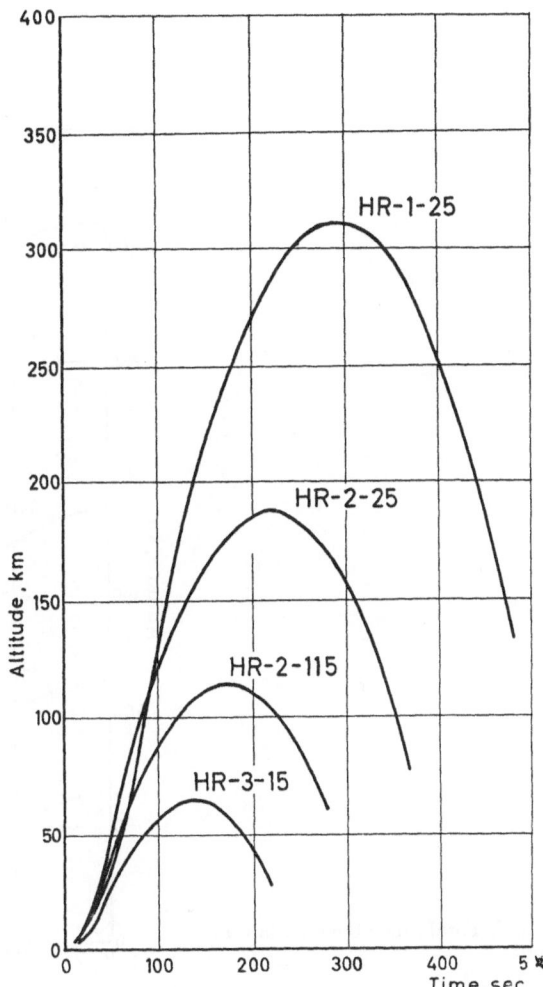

Fig. 6. HR-1, HR-2, and HR-3 altitude-time curves. Payloads 25 kg, 25—115 kg, and 15 kg respectively.

with time during the powered phases of HR-1 for a vertical launch with its largest payload 25 kg. The corresponding performance of the largest HR-3 (payload 15 kg) during the first 80 sec of flight is presented in Fig. 5.

Some altitude-time curves for different versions and payloads are shown in Fig. 6 and the total time above various altitudes in Fig. 7.

HR-1 will coast a 25 kg payload above 300 km altitude for 100 sec, HR-2 a 115 kg payload above 100 km for 100 sec, and the smaller HR-3 can bring 15 kg payload above 60 km altitude for 60 sec. Fig. 8 indicates the summit altitudes as a function of the payload.

Launching and Stabilizing

All Aurora rockets are unguided and spin-stabilized in order to make them simple and cheap. For the same reason and for the sake of mobility, it has been a design requirement that the launcher should be reasonably short. The two larger

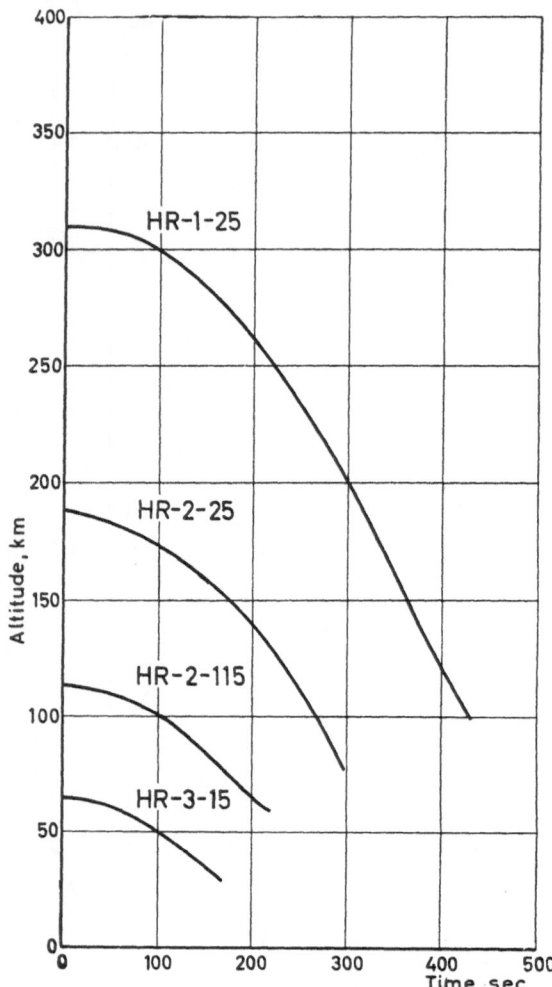

Fig. 7. Total time above various altitudes.

rockets, HR-1 and HR-2, have the same type of launcher with guide rails. For the smaller type HR-3, a simple zero-length launcher has been projected. (Fig. 9.)

The HR-1 and HR-2 launcher consists of an erecting rail mounted on a surplus 7.5 cm Bofors anti-aircraft gun carriage of the transportable type. After a slight modification of the gun carriage and its mechanisms for elevation and direction, it is suitable for mounting of the specially designed erecting rail.

The rail is about 8 m long and can be elevated manually or electrically to an angle of 70° to 90° at launching. When lowered in horizontal position, it can be folded in such a manner that it does not protrude outside the launcher carriage during transportation. Two longitudinal, slightly winding guide rails, attached to the erecting beam, provide guidance for the first 4 m of travel. Two shoes, fastened at the top and bottom of the rocket's first stage, fit into the guide rails and will give the rocket a spin rate of 0.7 rps when leaving the rails. The longitudinal velocity after launch (with the aid of booster rockets) is about 25 m/sec, a relatively low value when considering disturbances and gusts. However, the launching acceler-

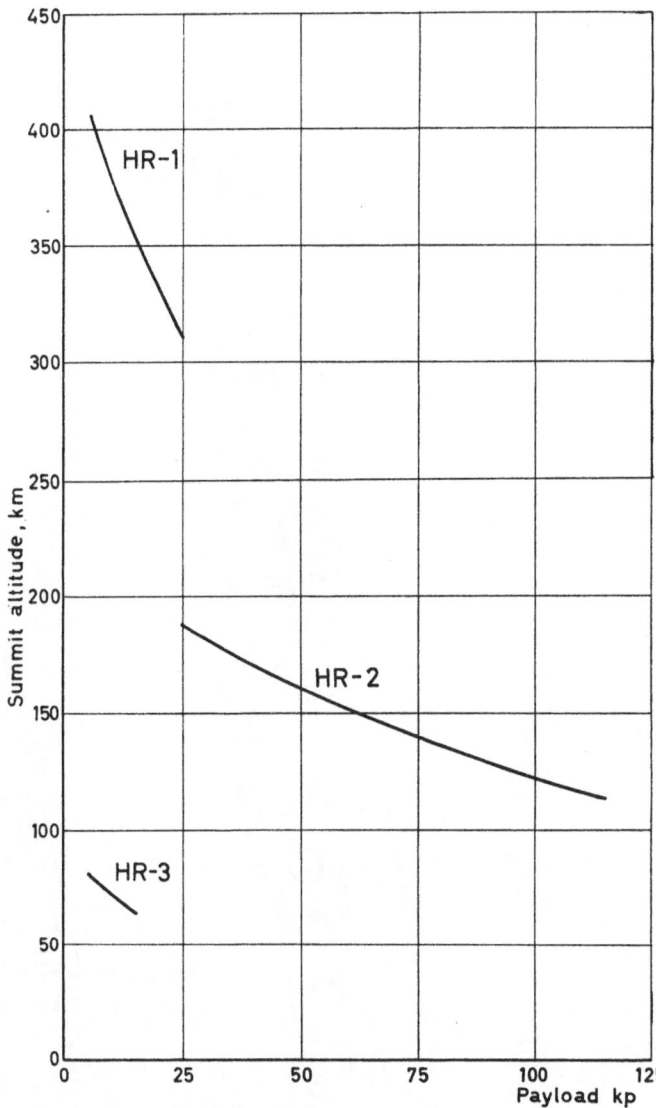

Fig. 8. Estimated vertical trajectories, summit altitude *versus* payload.

ation is about 7*g* and the trajectory precision is expected to be relatively good thanks to the combined effects of the following:

a. Axial spin, which will stabilize the motion against external disturbances and prevent trajectory dispersion due to geometrical imperfections (structural misalignment, excentrical thrust resultant).

b. Nose fins, loosely mounted at launching, to reduce dispersion due to wind gusts after launch.

The primary spin rate is maintained and increased to 1—1.5 rps with the aid of two exhaust pipes from the VR 3 turbine pump, arranged with tangential outlets in opposite directions at the base of the rocket, giving a maximum torque of about

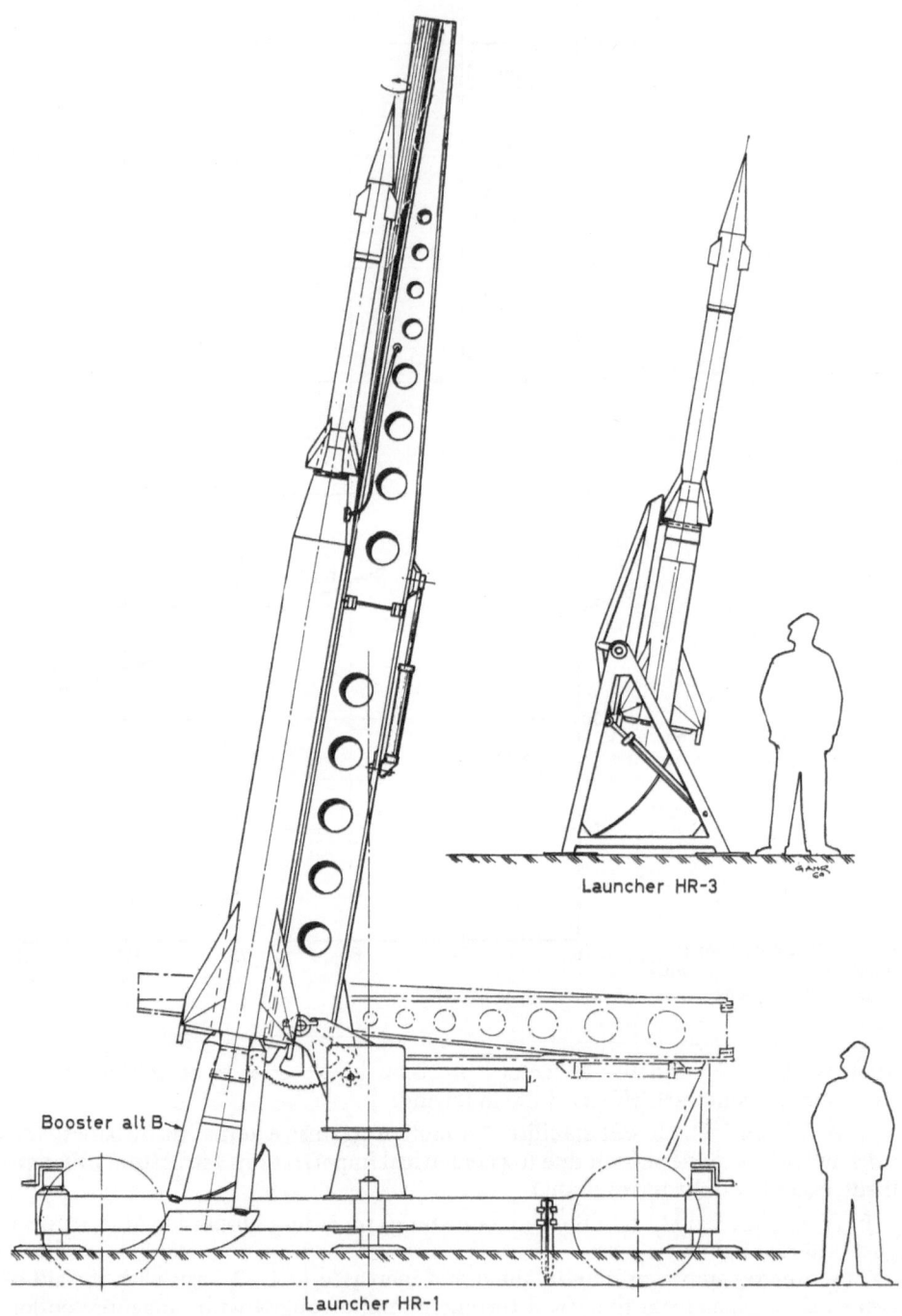

Fig. 9. Launchers for the HR-1 and HR-3 rockets.

FIXED BASE

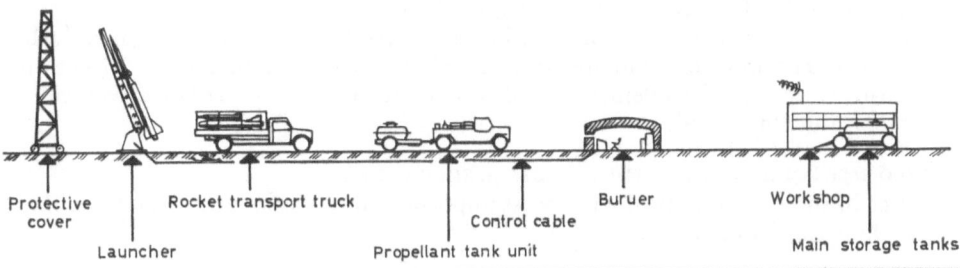

Protective cover   Rocket transport truck   Control cable   Buruer   Workshop

Launcher   Propellant tank unit   Main storage tanks

MOVABLE BASE

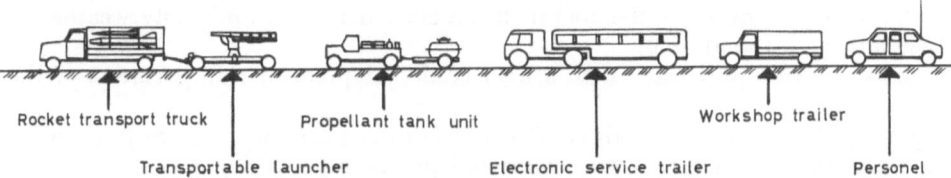

Rocket transport truck   Propellant tank unit   Workshop trailer

Transportable launcher   Electronic service trailer   Personel

Fig. 10. Ground equipment for HR-1 launching. Fixed and movable units.

5 kpm. After reaching a few kilometers altitude, where the speed is becoming quite high, the four tail fins will take over the spinning action with the aid of inclined rear trim tabs (about 8° inclined), and will maintain the rotation at about 1.5 rps when the rocket leaves the denser atmosphere. The spin rate of stage two will be kept equal to that of stage one by means of slightly inclined mounting of the four stabilizing fins.

The tendency of an unguided rocket with a large margin of stability to weathercock into the wind when the rocket has just left the launcher and the speed is low, can be partially eliminated by attaching the above mentioned dropable nose fins. For most high speed rockets, particularly those having liquid propellants, the center of gravity moves backwards at first during propellant expending and the center of pressure moves forward at increasing supersonic Mach number. By using dropable nose fins, the large static stability margin at launch can be temporarily reduced to such a level that the weathercocking action becomes non-critical. The nose fins are arranged to remain in position at low speed. At transsonic speed the dynamic pressure is sufficient to break them off, which is arranged to occur simultaneously. The stability requirement for high Mach number, high altitude flight can thus be fulfilled without penalty at launch.

The smaller rocket, type HR-3, has a zero-length launcher (Fig. 9) with direct release from two fixed points in the bottom stage. The spin (max. 1.5 rps) is first effected by slightly inclined booster nozzles. After separation, the spin of the second stage is maintained by inclined fins (same arrangement as stage 2 in the HR-1 version).

Ground Equipment and Personnel Organization

For launching it is assumed that much of the necessary equipment could be made available from existing facilities.

The necessary communication, meteorological, and launch area safety guard services should also be provided for through already established organizations, giving the research team their support during launching preparation and launching phases, as well as for payload recovery operations.

The launchings can take place either at a fixed main base or from movable launchers. The movable unit incorporates all necessary equipment for rocket and payload service and for telemetery and can be transported by road or by railway to suitable launching areas.

FIXED BASE EQUIPMENT FOR THE TWO STAGE ROCKET HR-1

Fig. 10 shows schematically the most important units which are needed at a fixed launching site for the HR-1.

Launch site equipment

1. Adjustable erecter-launcher, mounted on a concrete platform, with the same erector beam design as described above.

2. Protective cover for the rocket, to be erected in cold and gusty weather to facilitate service conditions.

3. Rocket transport truck with a light crane, adapted for transportation of the rocket stages.

4. Propellant tank unit, with HTP in a 500 litres trailer and a tractor jeep, which carries the kerosene tank and miscellaneous equipment.

5. Water tank at the launching site, with fire and drain hoses, plus standard chemical fire extinguishers.

6. A transportable electronic testing unit, for pre-launch checking of the rocket and payload circuits, *etc.*.

Additional main base equipment

1. Bunker for the launching crew interconnected with a launch control cable.

2. Workshops for the rocket and for electronic equipment, plus stores.

3. Main storage tanks for HTP and kerosene.

4. Rocket tracking ground stations, *e. g.* automatic optical teodolites, radio teodolites arranged on a suitable base line, Doppler radar for use with the rocket borne transponder.

5. A meteorological station, equipped with sounding balloons, *etc.*.

6. Telemetery receiving equipment and data processing unit.

7. Offices and laboratories for scientists and technicians.

8. Living quarters.

9. Transport vehicles for personnel and equipment.

MOVABLE BASE EQUIPMENT FOR ROCKET TYPE HR-1

The most important units are briefly described as follows (Fig. 10):

1. Transportable launcher, on a modified 7.5 cm Bofors anti-aircraft gun carriage.

2. Rocket transport truck, capable of loading two complete rockets HR-1 with separated second stages and miscellaneous other equipment. This truck is also used for towing the launcher.

3. Propellant transport unit (capacity for 2—3 firings).

4. Electronic service trailer. Fully equipped for payload preparation and for pre-launch checking of the rockets.

5. Transportable workshop.

6. Personnel transport automobiles.

For long distance transportation, the whole unit is preferably carried by railway.

Aurora — A Swedish High Altitude Rocket Project Study

It may be of interest to discuss briefly a typical personnel organization for the Aurora operations.

1. A project leader in charge of the entire project. The project leader heads the following sections and section chiefs.

2. Chief of launching operations, with crews for rocket preparation and launching, at least 6—8 technicians and mechanics.

3. Weather bureau.

4. Safety group.

5. Tracking, measuring and data processing section. Handles preparation and instrumentation payloads.

6. Transport section, payload recovery crew.

7. Quartering section.

Estimated Project Costs

Due to the limited design analysis carried out so far, it is difficult to estimate the development time and costs with reasonable accuracy. However, according to a preliminary estimate the following figures have been obtained.

Development schedule		HR-1	HR-3
Time from order to first experimental launching		1.5 years	1—1.5 years
Time to delivery of first production rocket		2.5 ,,	1.5—2 ,,
Development costs, rockets and launchers			
Development costs	Sw. crowns	2 million	0.8 million
	US $	400 000	160 000
Production costs			
Number of rockets produced		20	50
Production price per rocket	Sw. crowns	60 000	17 000
	US $	12 000	3 500
Total costs per rocket			
(excl. payload instrumentation)	Sw. crowns	160 000	33 000
	US $	32 000	7 000

Concluding Remarks

From this preliminary study it could be concluded that by utilizing already existing Swedish built rocket motors, it is possible to construct small or medium size sounding rockets apt for a variety of research missions. The overall costs including the development costs are, in spite of the comparatively small series assumed, not of such a magnitude that they exert a dominant influence on the total expenditure for a research program with sounding rockets.

It seems likely that the smaller and cheaper project HR-3, which with its maximum payload of 15 kg will reach a height of about 65 km, may be the most attractive alternative for starting a program. This project could easily be improved by arranging a better optimized combination of existing solid rocket motors giving heights around the 100 km mark.

This project study, which has been carried out as a voluntary enterprise within the Swedish Interplanetary Society (SIS), was initiated by Mr. Åke Hjertstrand, head of the Propulsion Section, who together with Mr. Lars Henrik Ågren, head of the Vehicle Section, undertook the preliminary investigation. When broadening the study, Mr. Olle Ljungström has played a major rôle.

LARS HENRIK ÅGREN, OLLE LJUNGSTRÖM and ÅKE HJERTSTRAND

In the enlarged study group the following members of the SIS have collaborated: Mr. Carl Gustav Ahremark, Mr. Göran Ehn, Mr. Kjell Flodin, Mr. Åke Hjertstrand, Mr. Åke Håborg, Mr. Gilbert Larsson, Mr. Olle Ljungström, Mr. Carl Reuterswärd, and Mr. Lars Henrik Ågren. Furthermore, Mr. Gillis Huss (Svenska Flygmotor Aktiebolaget) and Mr. Gert Larsson (Svenska Aeroplan Aktiebolaget) have contributed to the study.

References

[1] *Ankarswärd, B.*, VR 3 — Sweden's first pump fed liquid rocket engine. Astronautik *2*, p. 21 (1960).

[2] *Bisell, R. A.*, Performance characteristics of the two stage rocket Aeolus. Technical Note SAD 38. Wep. Res. Est., Australia.

[3] *Hjertstrand, Å.*, and *Ågren, L. H.*, Förslag till höghöjdsraket. Internal memorandum, May 20, 1959.

[4] *Newell, H. E.*, High Altitude Sounding Rockets Issue. Jet Propulsion *27*, p. 261 (1957).

[5] *Newell, H. E.*, Sounding Rockets. McGraw Hill, New York 1959.

[6] *Prout, P.*, and *Roth, C.*, Design criteria for upper atmosphere sounding rockets. ARS Preprint 813—59, June 1959.

[7] *Stroud, W. G.*, Meteorological rocket soundings in the Arctic. Jet Propulsion *28*, p. 817 (1958).

[8] Svenska Interplanetariska Sällskapet, Preliminär projektspecifikation, Höghöjdsraketsystemet "Aurora", February 5, 1960.

Small Sounding Rocket Propulsion Systems

John Gustavson

Grand Central Rocket Co., Redlands, Calif., USA

Very important contributions to the sciences of astronautics can be and are being provided by countries of limited funds, particularly in the field of sounding rocket experiments. The limited funding necessitates the optimization of the propulsion system. In some cases, the availability of inexpensive rocket motors makes it desirable to analyze how these rockets can be used with greatest success; in other cases a specific mission requirement of sending a payload to a desired altitude dictates the need for finding the best motor to do the job.

The present paper analyzes the effects of important propulsion system parameters on performance. The performance calculations have been made on relatively low cost electronic computers. The programming method is described in the appendix.

The propulsion system parameters include burning rate of propellant, density of propellant, and propellant mass fraction.

The optimization study in general analyzes the low cost, solid propellant propulsion system; however, a small part is devoted to the hybrid rocket system which combines the simplicity of the solid rocket with thrust control capability. Such hybrid rocket motors are adaptable to a very wide range of sounding rocket missions since the optimum thrust-time relationship can be obtained for the given mission thereby providing a standard "work-horse" propulsion system.

We have entered an era of intensified interest in upper atmosphere research, an era that demands increased efficiency and cost economy of sounding rocket systems. Universities and research groups that lack the large sums of money necessary for propulsion system development are seeking efficient sounding rocket systems that will provide a maximum return for funds expended. Countries not involved in solid propellant research for military uses are also interested in obtaining efficient sounding systems.

The majority of existing sounding rocket systems have been tailored to utilize already available solid propellant rocket motors. These motors are assembled into configurations capable of accomplishing the sounding objective — usually that of getting a payload of a certain weight to a desired altitude. The utilization of existing motors often lowers cost, since such motors are frequently available for little or no charge. But optimization of the system on the basis of performance is rarely achieved when existing motors are combined into a sounding rocket system.

The best approach to solving the problems of increasing efficiency and economy of sounding rocket systems would be to conduct an optimization study of propulsion systems.

A recent step in the right direction has been the development of certain end-burning rocket motors. However, the end-burning rocket motor is not necessarily the optimum propulsion system for sounding rockets. The end-burning rocket motor certainly possesses the advantage of a long burning time, thereby achieving low

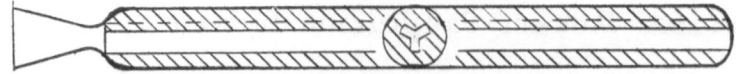

Fig. 1. Internal-burning rocket motor.

Fig. 2. End-burning rocket motor.

acceleration of payload, low aerodynamic drag, and minimum aerodynamic heating. On the other hand, major drawbacks of the end-burning motor are its low propellant mass fraction, resulting in decreased potential for heavy payloads, and its significant loss due to gravity.

This paper will show how the advantages of the internal-burning rocket motor may be optimized for sounding rocket purposes and how its disadvantages may be minimized.

Internal-Burning Rocket Motor

The basic components of an internal-burning rocket motor (Fig. 1) are a case, usually constructed of steel, aluminium, or fiberglass-wound plastic, and a nozzle, generally built of steel with graphite or other high temperature resistant inserts. The rocket case is insulated to protect it from the effects of the hot combustion gases. The entrance section to the nozzle also requires a certain amount of insulation. The rocket motor is loaded with a case-bonded propellant around a mandrel. Removal of the mandrel following propellant cure creates a center port extending the full length of the motor. The propellant burns outward from this center port perpendicularly to the case wall. The propellant itself shields the case against combustion gases, although in some designs propellant burns through to the case in some areas faster than in others.

An end-burning rocket motor (Fig. 2) consists of a case, protected during most of the burning by heavy insulation, and a nozzle. The end-burning motor is loaded almost completely with propellant. In contrast to the internal-burning motor which burns outwardly from the center port, the end-burning motor burns forward from its exposed aft end in cigarette fashion.

Table I presents a breakdown of typical weights for the two types of motors which have just been described. The table shows that the insulation forms an insignificant portion of the inert weight of the internal-burning motor. Insulation weight comprises a great bulk of inert weight in the end-burning motor. The table also shows clearly that the propellant mass fraction — the ratio of propellant available for combustion to the total rocket motor weight — is considerably greater for the internal-burning than for the end-burning motor.

TABLE I. Weight Distribution of Typical Solid Rocket Components

Rocket Motor	Internal-Burning (percent)	End-Burning (percent)
Case weight	17	18
Nozzle weight	4	6
Insulation weight	2	12
Total inert weight	23	36
Propellant weight	77	64
Motor weight	100	100

Where the propellant weight of the two motors is equal, one would immediately conclude that the internal-burning motor offers the greater potential performance. This would be true if it were not for the limiting factors of acceleration on the payload, aerodynamic drag, and aerodynamic heating.

The burning time of both types of motors is determined by the web thickness of the motor grain. The internal-burning rocket motor has a relatively thin web, while the end-burner has an extremely long web. Depending upon the burning rate of the propellant, the burning time of both motors can vary within certain limits. However, it may easily be seen that the end-burner will always be capable of providing a much longer burning time than the internal-burning rocket motor. If the two motors are loaded with the same amount of propellant, the same total impulse will be available from both motors. To provide the same total impulse yet widely differing burning times for the two motors, naturally creates two different thrust levels. If we assume for the sake of simplicity that a constant thrust level is maintained, the internal-burning motor will burn at low thrust over an extended time period.

For reasons of flight stability, the initial acceleration of the rocket motor is important. If the rocket does not attain a high flight velocity immediately after launch it will not have sufficient stability to follow the planned trajectory. Therefore, the high thrust-to-weight ratio of the internal-burning motor offers the most favorable conditions for achievement of the trajectory immediately after take-off. This high thrust level, however, is detrimental in the final burning phases when the payload experiences very high accelerations because of the high thrust-to-weight ratio. An end-burning motor is therefore the most satisfactory in the later phases of burning. Many attempts have been made to combine these two desirable features, that is to incorporate the high initial acceleration of the internal-burning motor and the low final acceleration of the end-burning motor. In most cases, however, the difference in thrust level is so great that it becomes inefficient from a structural standpoint to incorporate the two thrust levels into one motor. Instead, a booster of the internal-burning type may be used to boost an end-burning rocket. The result is a more complex system with increased cost and less reliability.

A better solution would be to bridge the gap between the long burning time and low thrust level of the end-burner and the short burning time and high thrust level of the internal-burning motor. The performance of either type of motor can be improved to bridge the gap.

Aerodynamic Effects

As mentioned previously, the high acceleration of the internal-burning motor will give the rocket system its maximum velocity at a rather low altitude. Maximum velocity in the denser part of the atmosphere will result in high values of aerodynamic drag and aerodynamic heating on the nose cone, payload, and in some instances on the case wall. These limitations do not apply to the end-burning rocket, which accelerates at a low rate until burnout at the fringe of the atmosphere. Detrimental features of the end-burning motor are high trajectory dispersion due to the long burning time and low aerodynamic stability during flight, and the high loss to gravitation produced by the long burning time.

This paper will not attempt to describe how an end-burning rocket may be modified to offer higher performance, lower trajectory dispersion, and simpler take-off characteristics. Instead, we will concentrate on how to improve the performance of the internal-burning rocket sounding system.

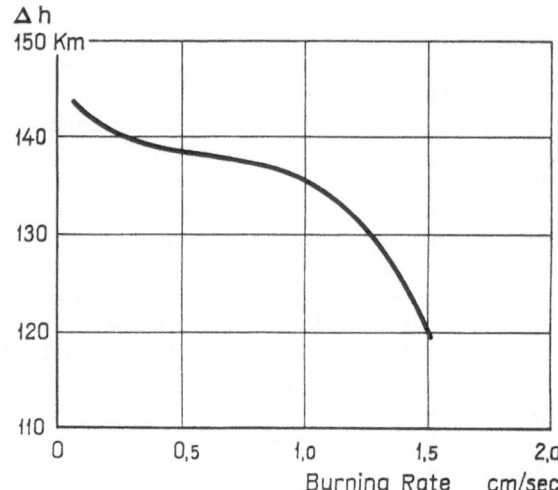

Fig. 3. Peak altitude *versus* burning rate.

Burning Time Increase

Although some payloads can be designed today to withstand accelerations in excess of 100 *g* and would therefore be excellently suited for internal-burning rocket motor boosting, it still remains an important task of the rocket designer to design motors with lower acceleration in the final phases of flight and lower velocity during passage through the sensible atmosphere. To meet the need, very slow burning solid propellants are being developed for internal-burning rocket motors. The standard burning rate for solid rocket propellant is about one cm/sec. This value provides burning times of about two to four seconds for the standard types of sounding rocket motors. The significant increase in peak altitude achieved by use of slower burning propellants is illustrated in Fig. 3. (See Appendix for Motor Parameter Calculations and Trajectory Calculations.)

The peak altitude reached by a sounding rocket increases as the burning rate of the propellant is decreased. This is the result not only of the lower drag effect but also because of the higher propellant mass fraction obtainable with low burning rate propellant. The nozzle throat area decreases in size as the mass flow rate of the combustion gases is decreased.

Propellant burning rate can be reduced somewhat by changing the grind of oxidizer to coarser particles. In addition, various burning rate depressants may be introduced into the propellant without significantly altering its specific impulse. We are hopeful that such roads of research will eventually lead to development of propellants with very slow burning rates and long burning times, furnishing internal-burning rockets with some of the advantages of the end-burner motor.

Besides changing the composition of the propellant to provide slower burning time, the pressure of the motor can be lowered. This, however, can only be done within certain limits because of the corresponding detrimental effect on the thrust coefficient, particularly at sea level altitude.

High Density Propellants

Another step which can be taken to decrease drag and improve rocket performance is the employment of high density propellants. Standard propellants have densities of the order 1.75 g/cm³. Propellant densities in excess of 2 g/cm³ can be obtained, however. If a higher density propellant is produced which has approxi-

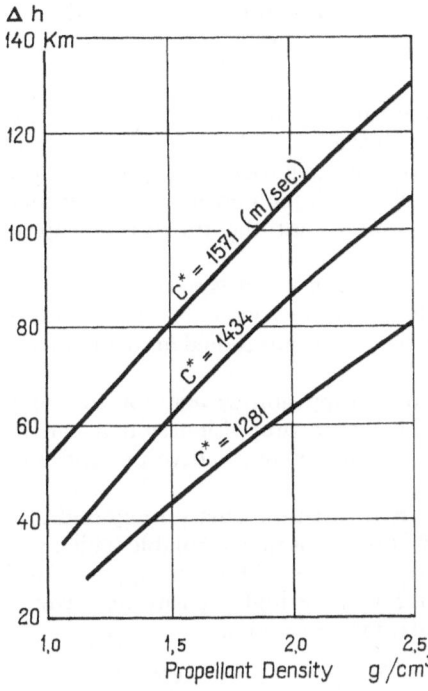

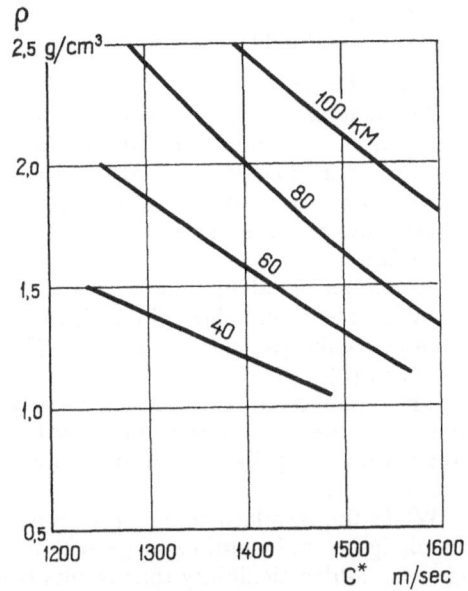

Fig. 4 (left). Peak altitude *versus* propellant density.

Fig. 5 (right). Trade-off between density and characteristic exhaust velocity for several peak altitudes.

mately the same energy content as standard density propellants, a greater performance can be obtained for the internal-burning rocket motor.

This improved performance is obtainable because the high density propellant makes it possible to decrease the motor diameter to contain the same weight of propellant as when standard density propellant was used. Consequently, the rocket has a lower frontal area drag or a lower area-to-weight ratio. The improved performance also results from the fact that the rocket motor envelope has become smaller than was needed for the standard density propellant. The rocket motor envelope now weighs less while the propellant retains the same weight as the standard density propellant in the larger diameter case. The result is a higher propellant mass fraction. The combination of the more favorable frontal area-to-weight ratio and the higher propellant mass fraction is shown in Fig. 4.

There are two basic methods of obtaining higher density of a propellant. One method is to incorporate high density inert components in the propellant which increase density, yet because of the inertness of the additives simultaneously decrease specific impulse. This method, which has been studied by Seifert [1], employs metal additives such as lead or tungsten.

The second method is to include components in the propellant which raise density while maintaining the specific impulse at practically the same original level. This method may employ the use of additives such as heavy metal hydrides. The basic heavy metall element increases density while the hydrogen decreases the molecular weight of the exhaust gas, thereby maintaining a high specific impulse. In some instances, the metals or metal hydrides have a high combustion heat which aids in retaining the favorable specific impulses.

Fig. 5 shows trade-offs between density and characteristic exhaust velocity for certain representative peak altitudes. It appears clearly that significant gains can be achieved by increasing the propellant density, even if the specific impulse decreases accordingly.

Varying Thrust

The calculations in the previous section were based on constant pressure-time and roughly constant thrust-time curves. The trajectories did take into account changing thrust coefficient with altitude, but for sounding rockets designed to work at high pressure levels the change from sea level to altitude performance is not too significant. We are therefore justified in assuming that the thrust-time curves were flat.

A number of calculations were run in which the pressure was varied during the burning. The results showed that it would be desirable to vary the thrust to optimize for a particular payload and altitude. This problem has been studied mathematically by Tsien [2].

The greatest advantage of a variable thrust propulsion system for sounding rockets is the flexibility of usage achieved. The same propulsion system can be utilized under optimum conditions over a wide range of payload values and altitudes.

While the standard solid rocket motor has its thrust-time curve integrated into the design from inception, another type of rocket motor is now available which will give the added flexibility that results from varying thrust. This motor, which will be described in the following section, also offers very high density and would therefore have low drag and rather high propellant mass fraction.

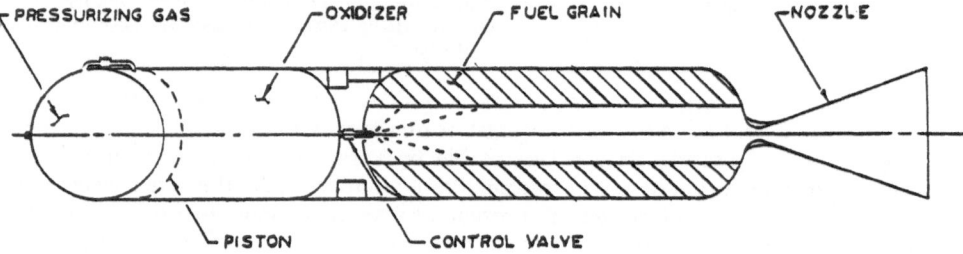

Fig. 6. Hybrid rocket motor.

The Hybrid Rocket Motor

The general layout of a hybrid rocket motor propulsion system [3] is shown in Fig. 6. A case-bonded, solid fuel chamber is equipped with a nozzle in standard fashion. A liquid oxidizer is forced into the solid fuel lined chamber by means of a gas pressurizing system. The hybrid rocket motor is therefore not a pure solid propellant rocket motor, since it requires a liquid oxidizer in its operation. However, it has most of the advantages of the solid rocket motor plus the added advantage of thrust level modulation and complete on-off control. The on-off control and re-start capability is not quite as important in sounding rocket applications as in other applications. Table II outlines the important parameters of the hybrid rocket propulsion system and provides an example of the fuel and oxidizer combination. As

TABLE II. Hybrid Rocket Motor Propellant and Parameters

Oxidizer	Chlorine Trifluoride (ClF_3)
Fuel	Aluminized Rubber (Al-PBAA)
Specific Impulse*	272 sec
Specific Density Impulse	510 g sec/cm³

* Theoretical, shifting equilibrium, expansion from 50 atmospheres to sea level.

will be noted, such a system has both high density and high specific impulse. The hybrid rocket motor is controlled by throttling the liquid injector valve. This one moving part is all that is needed to give a pressure-time curve which best suits the flight trajectory desired. High takeoff acceleration can be obtained, thereby resulting in initial flight stability, followed by low thrust flight through the denser part of the atmosphere. Finally, somewhat higher but still regressive thrust during the concluding phases before burnout will accelerate the rocket to high burnout velocities.

The propellants for hybrid rocket motors utilize the same oxidizer as storable liquid systems. Consequently, they enjoy the same advantages as liquid systems. In addition, the hybrid motor propellant can contain high density or low molecular weight, metal hydrides yielding high density impulse and high specific impulse. The hybrid rocket motor chemicals are readily obtainable and very safe in storage. The physical properties of the solid grain are outstanding since the solid loading in the rubber binder system can be kept low without performance loss. Temperature cycling should also present less of a problem for hybrid rocket motors. Propellant limitations are set by the freezing point of the liquid oxidizer rather than by the physical properties of the solid fuel grain. The propellant combinations for the hybrid rocket motor should be hypergolic to provide on-off operation without the use of a conventional igniter. By presetting the oxidizer flow valve any constant thrust-time relationship can be obtained. This way drag and acceleration requirements can be met easily. The hybrid motor should therefore see increased use for sounding rocket applications in the future.

References

[1] *Seifert, Howard S.*, Effect of Varying Propellant Density on Rocket Performance, Journal of American Rocket Society 22, 213 (1952).
[2] *Tsien, H. S.* and *Evans, R. C.*, Optimum Thrust Programming for a Sounding Rocket, Journal of American Rocket Society, *21*, 99 (1951).
[3] *Gustavson, J.*, The Hybrid Rocket Motor and Its Unique Capabilities, ARS preprint 1167-60.
[4] *Thackwell, Jr., H. L.*, Generalized Study of Solid Propellant Rockets for Upper Stages of Multistage Rocket Vehicles, 16th JANAF meeting, March 1, 1960.

Appendix

Motor Parameters

The components of an internal-burning rocket motor system, based on derivations by Thackwell [4], can be estimated from the following equations for the weights of the propellant, the case, the insulation, and the nozzle.

Propellant Weight:

$$W_p = \pi d^3 \left[\frac{4}{\varrho_p (l/d + 2/3)} + \frac{8C^* a (A_p/A_t)}{k\, g\, p^{1-n}} \right]^{-1} \tag{1}$$

where k is the web fraction of the motor, *e.g.*, $k = 2w/d$

Case Weight: $$W_c = \frac{\pi d^3}{2}\, k_2 \frac{\varrho_c}{F_{ty}}\, p(l/d + k_1) \tag{2}$$

where k_1 is a case geometry constant and k_2 summarizes the pressure variation and the safety factor.

Insulation Weight: $W_1 = \pi d^3 k_3 (l/d + 3)$ (3)

where k_3 is a constant depending on the particular grain design.

Nozzle Weight: $W_n = k_4 \dfrac{C^* C_F}{g} W_p$ (4)

which basically expresses the nozzle weight as a function of total impulse and a constant k_4 which is dependent on the burning time, the chamber pressure, and the motor diameter.

The representative values shown in Table III were used for the motor calculations. As soon as a value for the propellant mass fraction was obtained, it was entered in the computer for a performance run, as described below.

Table III. Motor Values Employed in Study

W_i	5 kg	l/d	15
W_p	45 kg	d	14.37 cm
W_m	50 kg	ε	15
$W_{payload}$	25 kg	k	0.3
W_s	7 kg	k_1	1.0
W_o	82 kg	k_2	1.5
C^*	1570 m/sec	k_3	0.001 g/cm³
P_c	50 atm	k_4	varies from 0.4 to $1.0 \cdot 10^{-4}$ sec⁻¹

Trajectory Calculations

The vertical or near vertical trajectories for an internal-burning sounding rocket system were computed on a Bendix G-15. This program is illustrated in Fig. 7. It employs the step-by-step integration method, and takes into account pressure changes in the rocket motor as well as pressure changes of the atmosphere outside the rocket motor. The program also corrected for overexpanded nozzles, see Eq. (6). It was assumed that separation will take place in the nozzle and create near-optimum expansion conditions at the point of separation. In that case the exit plane pressure will equal the ambient pressure. For conditions where the ambient pressure is lower than exit pressure, Eq. (5) is used. These conditions exist at ideal or under-expanded conditions such as are found in the upper atmosphere. The use of the chamber pressure for calculations of thrust makes it possible to calculate more accurate trajectories. The propellant flow rate is also expressed as a function of pressure in Eq. (8) for the calculation of the instantaneous weight of the rocket.

Eq. (9) is the incremental velocity equation, the angle ϑ being measured between the local vertical and the direction of flight of the rocket. The equation for ϑ was calculated on the basis of the gravity turn, a turn where the thrust vector of the motor at all times is parallel to and in the same direction as the velocity vector of the rocket motor.

The program used average conditions to calculate each increment, and after iteration a value of high accuracy was found. The accuracy of this trajectory is about one percent for the burnout velocity and about two percent for the angle. The equations follow below.

$$F_n = A_t P_c C_{dis} \left\{ \sqrt{\frac{2\gamma^2}{\gamma-1} \left(\frac{2}{\gamma+1}\right)^{\frac{\gamma+1}{\gamma-1}} \left[1 - (N)^{\frac{\gamma-1}{\gamma}}\right]} + \left(N\varepsilon - \frac{P_a \varepsilon}{P_c}\right) \right\} \quad (5)$$

$$\text{if } \frac{P_a}{P_c} < N$$

$$F_n = A_t P_c C_{dis} \sqrt{\frac{2\gamma^2}{\gamma-1} \left(\frac{2}{\gamma+1}\right)^{\frac{\gamma+1}{\gamma-1}} \left[1 - \left(\frac{P_a}{P_c}\right)^{\frac{\gamma-1}{\gamma}}\right]} \qquad \text{if } \frac{P_a}{P_c} \geqq N \quad (6)$$

where A_t, C_{dis}, γ, N, and ε are constant throughout the problem, P_c is obtained from motor pressure data at time n, P_a is calculated from h_{av}.

$$D_n = \tfrac{1}{2} S C_D (v_{av})^2 \varrho \qquad (7)$$

where S is a constant for the rocket, C_D is obtained from drag curves as function of M at time n, ϱ is calculated from h_{av}, and v is the average between v_n and v_{n-1}.

$$W_n = W_{n-1} - P_c A_t \Delta t \, g / C^* \qquad (8)$$

where g, A_t, C^* are constant, $\Delta t = t_n - t_{n-1}$, P_a is at time n.

$$v_n = v_{n-1} + g \Delta t \left[\frac{F_n - D_n}{W_n} - \left(\frac{R_0}{R_0 + h_n}\right)^2 \cos \vartheta_n \right] \qquad (9)$$

$$h_n = h_{n-1} + v_{av} \cos \vartheta_n \Delta t \qquad (10)$$

$$\vartheta_n = \vartheta_{n-1} + 57.3 \sin \vartheta_n \Delta t \, g / v_{av} \qquad (11)$$

Nomenclature

a	acceleration	S	cross-sectional area
A_p/A_t	port-to-throat area ratio	t	time
C^*	propellant characteristic velocity	v	velocity
C_D	drag coefficient	w	web thickness
C_{dis}	discharge	W	weight
C_F	nozzle thrust coefficient	γ	specific heat ratio
d	diameter of motor	ε	nozzle area expansion ratio
D	drag	ϱ	density
F	thrust	ϑ	angle of flight from local vertical
F_{ty}	allowable yield stress of case material		
g	acceleration due to gravity		*Subscripts*
h	altitude	a	ambient
k	constant	c	case or chamber
l	length of motor	i	insulation
M	Mach number	m	motor
N	P_e/P_c for given expansion ratio	n	nozzle
P	pressure	o	initial
r	burning rate	p	propellant
R_0	radius of earth	s	structures, fins

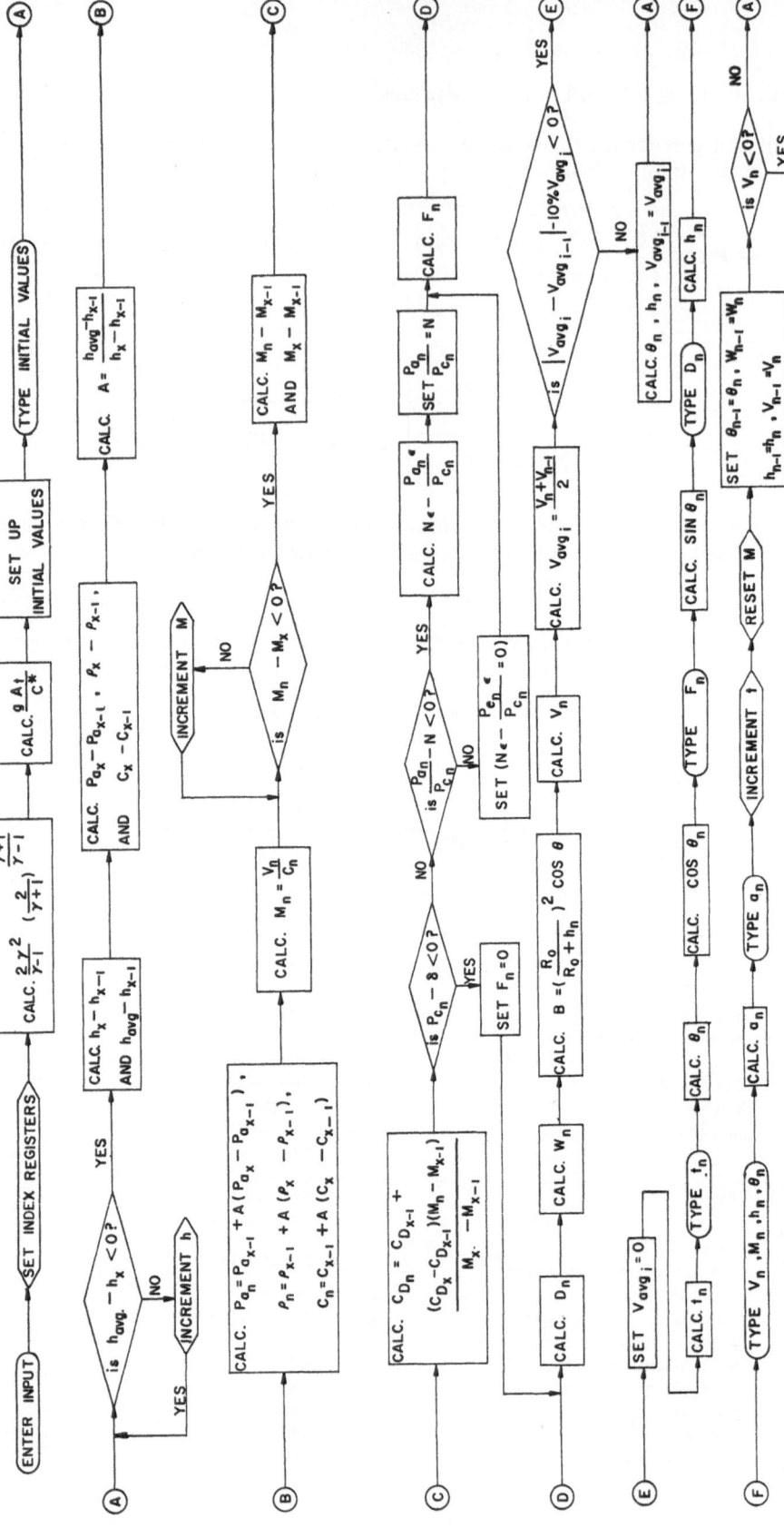

Fig. 7. Trajectory flow chart.

Influence of the Intermediate Flight Time on the Performance of Multistaged Sounding Rockets

Glauco Partel and Rolf Engel

Società Italiana Sviluppo Propulsione a Reazione (SISPRE), Rome, Italy

For a long time in Italy the problem of sounding rockets has been studied. If you have a multi-stage rocket, there are two ways to follow in the flight procedure:
(1) to ignite every step one after each other;
(2) to delay, for instance, the ignition of the second stage for a certain time during which the vehicle is coasting. This coasting flight can be done together with a burn-out first stage or after separation from the burn-out first stage.
It is known that a period of coasting flight normally gives a higher peak altitude.

SISPRE was charged with the study of this problem theoretically and with particular consideration to multi-stage rockets realized with existing propulsion units available immediately.

Besides this, SISPRE has a task to develop relatively small two-stage sounding rockets, at present being tested in the Italian proving ground of Sardinia. These rockets serve, among other purposes, to study the reliability of stage separation and to check and improve all existing systems of telemetering and observation during flight.

Results of Studies

It is not possible to give results of all the vast amounts of studies we have carried out; therefore, we choose only some indicative examples. Naturally, it is possible to study the problem in question by the calculus of variations, but we found that the method of direct variations of the parameters is most speedy for the electronic computer.

The following configurations were analyzed:

(a) Three-stage vehicles:
    (1) SEPR 732 — ASP 1 — C 7 (BPD)
    (2) NIKE — ASP 1 — C 7 (BPD)
    (3) SEPR 732 — SEPR 705 — C 7 (BPD)
    (4) SEPR 732 — Cluster of 3 C 7 — C 7 (BPD)
    (5) SEPR 732 — SEPR 5053 — C 7 (BPD)

(b) Two-stage vehicles:
    (6) ASP 1 — C 7 (BPD)
    (7) SEPR 705 — C 7 (BPD)
    (8) SEPR 5053 — C 7 (BPD)
    (9) ASP 3 — C 7 (BPD)

The choice of these vehicles was dictated only by reasons of price and availability. The data of these rockets are well known from the existing literature and for this

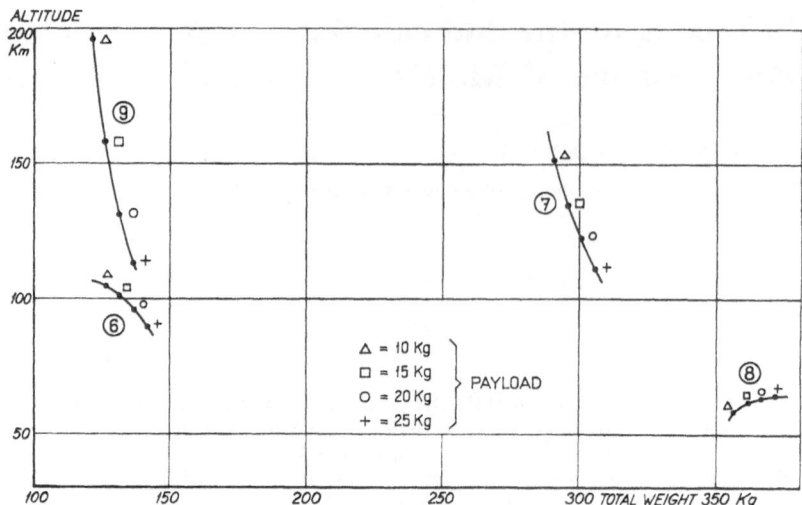

Fig. 1. Two Stage Sounding Rockets.

reason are not repeated here. C 7 is a solid propellant rocket developed by the Italian firm BPD, and has the same characteristics as the US Falcon engine (18 kg propellant and 12 kg hardware).

First consideration is the total weight of vehicle, which gives an indication of the total price, an important factor for the customer.

Each of the above mentioned vehicles has been considered with a payload of 10, 15, 20, and 25 kg.

For every configuration the drag curve was calculated as a function of Mach number and Reynolds number with and without jet. The trajectories were computed by Prof. Dr. Bödewadt with the aid of the Ferranti digital electronic computer of the Italian National Council of Research. The data for the atmosphere were taken from the Standard Atmosphere ARDC.

Fig. 1 gives the dependence of the peak altitude as a function of the total weight with intermediate flight time $t_1 = 0$ sec. The two-stage configuration ASP 3 — C 7 is clearly superior in altitude and price.

Fig. 2 gives the relationship for the three-stage vehicles, also with intermediate flight time $t_1 = 0$ sec. It is clear that if a minimum peak altitude of 500 km is desirable, only vehicles (1) or (3) are of interest.

Now, the whole picture changes very much if we carry out the same calculations with an intermediate flight time, *i.e.* an optimization of the curve.

Fig. 3 gives the picture for the two-step configuration (9). For the different payloads, the altitude is given as a function of intermediate flight time. The small diagram inserted in this figure gives the corresponding maximum altitude as a function of the payload. With approximately 20 sec intermediate flight time, it is possible for a small payload to double the altitude.

Fig. 4 gives for the different payloads the peak altitude as a function of the intermediate flight time. This case considers the separation of the first stage to be carried out immediately after its burn-out and that only the second and third stages are coasting in free flight. In this case it is clear that the most efficient procedure is to fly without any intermediate flight time.

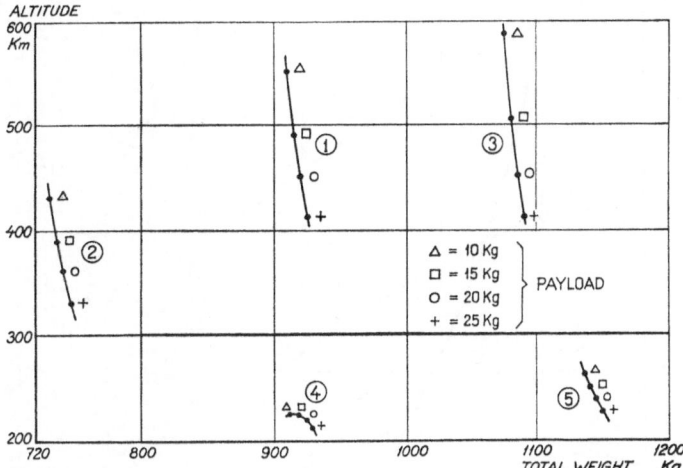

Fig. 2. Three-Stage
Sounding Rockets.

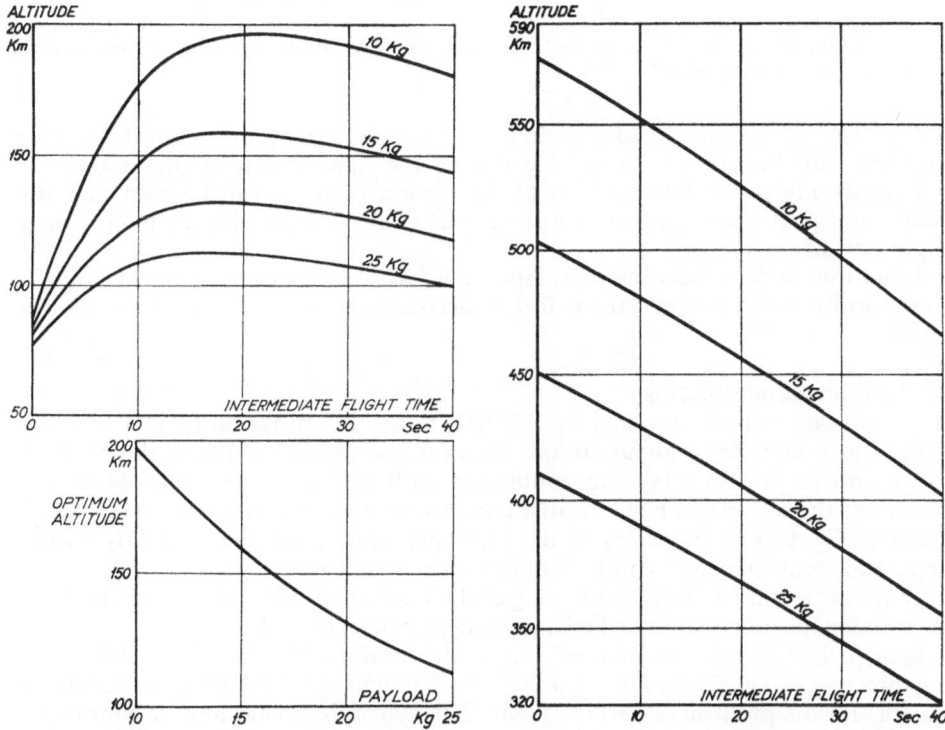

Fig. 3. ASP 3—C7 (configuration 9).

Fig. 4. SEPR732—SEPR705—C7 (configuration 3). Coasting without booster.

65

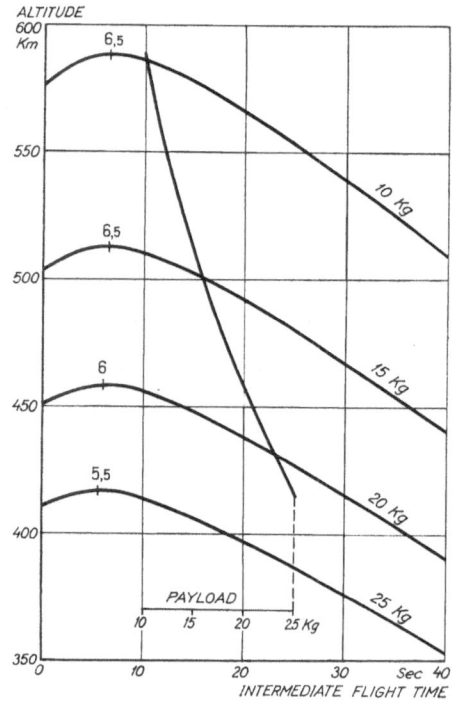

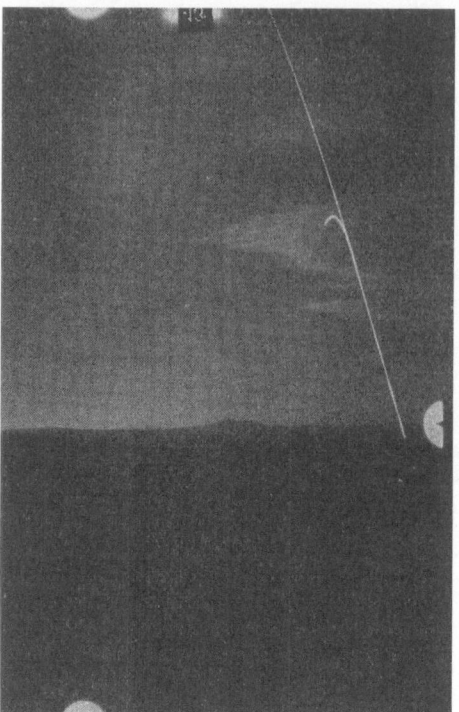

Fig. 5. SEPR732—SEPR705—C7 (configuration 3). Coasting with booster.

Fig. 6. Separation of two-stage test vehicle.

Fig. 5 gives the same configuration, but here the empty first stage is coasting together with the rest of the vehicle during the intermediate flight time and is separated only just before ignition of the second stage. It can be seen that this procedure gives some gain of altitude for a small payload with an intermediate flight time of about 6 sec.

From this study we can see that, especially for three-stage configurations, a vast theoretically study is necessary before a decision about the final vehicle can be taken.

SISPRE Sounding Rockets

A two-stage vehicle, realized by SISPRE under the direction of Dr. Angeloni with a peak altitude of about 30 km, has been successfully tested in the Sardinia range and at present trials are continuing with different separation devices. A picture of the separation between first and second stage, taken by cinetheodolite, is given in Fig. 6. The trajectory of the first stage after separation is clearly visible. Separation took place at an altitude of approximately 1000 meters.

In the same time, tests for recovering of the payload are also under way and serve for training the observation and telemetering personnel as well.

Besides the Cubic installation existing in the Sardinia test range, SISPRE under the direction of Dr. Fantoni has studied a new system of telemetering, particularily suitable for the problem of vertical flight. This system is mobile, and its handling is easy since it should be used also for future meteorological rockets.

Studies are at present continuing in SISPRE for a three-stage sounding rocket.

On the Dynamics of a Space Vehicle, Equipped with One Main Rocket Motor and Two Vernier Motors

J. M. J. Kooy

Royal Military Academy, Breda, the Netherlands

The general theory of the rotary motion of a rocket vehicle as indicated is developed. The main rocket motor is mounted in such a way that a required pitch moment and yaw moment of the main thrust can be obtained. The function of the Vernier motors is only to provide a roll moment, if required.

The general problem to be discussed is, how to determine numerically the motion of the vehicle, if the two angles α and β, defining the relative direction of the main thrust and the angle γ, defining the adjustment of the pair of Vernier motors, as well as the magnitudes of the three thrusts are prescribed as functions of time. Thereby also the influences of the Coriolis accelerations of the fluid particles in the nozzles and the gyroscopic action of the rotating turbo pump assembly for supplying the fuel to the rocket motors, are taken into account, as well as the aerodynamic moments coming into play.

In this article the general dynamic theory is developed of a rocket vehicle, equipped with one main rocket motor and two Vernier motors.

For steering the main rocket motor is mounted in such a way that the angle included by the central line of the main nozzle and the (coplanar) roll axis of the vehicle can be varied, whereas the plane containing these two axes can be varied as to the rocket frame. (Hence the position of the main rocket engine as to the rocket frame is determined by two angles α and β.) The two Vernier motors are situated symmetrically on both sides of the main rocket motor and are both swivel mounted. In mid position the central lines of the nozzles of the three motors are parallel with the roll axis of the vehicle and in one plane whereas the axis of the main thrust tube then coincides with the roll axis. The function of the Vernier motors is only to provide a roll moment (about the roll axis), if this will be required. Therefore any Vernier motor can turn with the nozzle axis in a plane perpendicular to the plane of the three nozzle axes in mid position. If then one Vernier motor has a deflection γ, the other Vernier motor has a corresponding deflection $-\gamma$. The required steering moments about the pitch axis and the yaw axis of the vehicle are obtained by suitable adjustment of the direction of the thrust of the main rocket motor with respect to the vehicle frame.

In world space the rocket vehicle may be supposed to move in a resultant field of force (being in general a combined gravitational and inertial field) due to different celestial bodies. In case of the ascending powered flight through the terrestrial atmosphere, also aerodynamic forces and corresponding moments about pitch axis and yaw axis come into play. As to these aerodynamic forces and corresponding moments, the simplifying supposition is made that the external form of the vehicle has rotational symmetry as to the roll axis.

The general problem considered in this article is, how to determine numerically the motion of the vehicle, if the two angles α and β, defining the relative direction of the main thrust and the angle γ, defining the adjustment of the pair of Vernier motors, as well as the magnitudes of the three thrusts are prescribed as functions of time. Thereby also the influence of the Coriolis accelerations of the fluid particles in the thrust tubes and the gyroscopic action of the rotating turbo feeding pump assembly for supplying the fuel to the rocket motors is taken into account.

1. Coordinate systems used

The following coordinate systems and corresponding relations are used:

$x\,y\,z \rightarrow$ rectangular geocentric system of reference which does not rotate with respect to the celestial sky. The origin $x = 0$, $y = 0$, $z = 0$ coincides with the centre of the earth. The unit vectors in $x+$, $y+$ and $z+$ direction will be denoted by $\boldsymbol{i}, \boldsymbol{j}, \boldsymbol{k}$.

$X\,Y\,Z \rightarrow$ vehicle rigid system of reference with origin at centre of gravity of vehicle + still included fuel and Z-axis coinciding with roll axis of vehicle.

$\boldsymbol{i}_1, \boldsymbol{j}_1$ and $\boldsymbol{k}_1$ are unit vectors in $X+$, $Y+$ and $Z+$ direction. $\boldsymbol{k}_1$ is pointing to vehicle nose.

$\mathsf{x\,y\,z} \rightarrow$ other vehicle rigid system of reference, with corresponding unit vectors $\mathbf{i}, \mathbf{j}$ and $\mathbf{k}$.

$\underline{x}\,\underline{y}\,\underline{z} \rightarrow$ coordinate system, parallel with $x\,y\,z$, with origin at centre of gravity of vehicle + still included fuel.

The transference from $\underline{x}\,\underline{y}\,\underline{z}$ to $X\,Y\,Z$ can be carried out by three consecutive rotations, about angles φ, ϑ and ψ:

$$\underline{x}\,\underline{y}\,\underline{z} \rightarrow X'\,Y'\,\underline{z} \rightarrow X''\,Y'\,Z \rightarrow X\,Y\,Z$$

$$\underbrace{\hphantom{xxxx}}\qquad \underbrace{\hphantom{xxxx}}\qquad \underbrace{\hphantom{xxxx}}$$

$$\varphi \text{ about } \underline{z} \qquad \vartheta \text{ about } Y' \qquad \psi \text{ about } Z$$

The rotations φ, ϑ and ψ are in right handed relation to $\underline{z}+$, $Y'+$ and $Z+$. Hence $X'\,Y'\,\underline{z}$ and $X''\,Y'\,Z$ are intermediate systems of reference.

In Fig. 1, the axis X' is the projection of the axis X'' on the plane $\underline{x}\,\underline{y}$. From Fig. 1 we read:

$$\cos(X''\underline{x}) = \cos\vartheta\,\cos\varphi$$
$$\cos(X''\underline{y}) = \cos\vartheta\,\sin\varphi \tag{1}$$
$$\cos(X''\underline{z}) = -\sin\vartheta$$

$$\cos(Y'\,\underline{x}) = -\sin\varphi$$
$$\cos(Y'\,\underline{y}) = \cos\varphi \tag{2}$$
$$\cos(Y'\,\underline{z}) = 0$$

$$\cos(Z\,\underline{x}) = \sin\vartheta\,\cos\varphi$$
$$\cos(Z\,\underline{y}) = \sin\vartheta\,\sin\varphi \tag{3}$$
$$\cos(Z\,\underline{z}) = \cos\vartheta$$

Hence:

$$X'' = \underline{x}\cos\vartheta\,\cos\varphi + \underline{y}\cos\vartheta\,\sin\varphi - \underline{z}\sin\vartheta$$
$$Y' = -\underline{x}\sin\varphi + \underline{y}\cos\varphi \tag{4}$$
$$Z = \underline{x}\sin\vartheta\,\cos\varphi + \underline{y}\sin\vartheta\,\sin\varphi + \underline{z}\cos\vartheta$$

On the Dynamics of a Space Vehicle

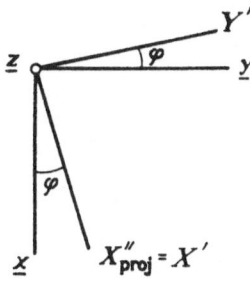

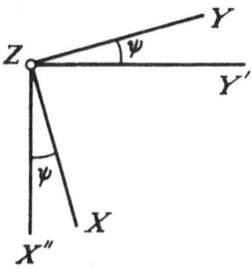

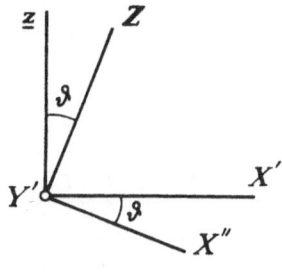

Fig. 1. $\underline{z}$ axis $\perp$ to drawing-plane, pointing to observer.

Fig. 2. Z axis $\perp$ to drawing-plane, pointing to observer.

Fig. 3. Y' axis $\perp$ to drawing-plane, pointing from observer.

Further we read from Fig. 2:

$$X = X'' \cos\psi + Y' \sin\psi$$
$$Y = -X'' \sin\psi + Y' \cos\psi \tag{5}$$

Substituting (4) in (5) we obtain:

$$X = \underline{x}\,(\cos\vartheta\,\cos\varphi\,\cos\psi - \sin\varphi\,\sin\psi) +$$
$$+ \underline{y}\,(\cos\vartheta\,\sin\varphi\,\cos\psi + \cos\varphi\,\sin\psi) + \underline{z}\,(-\sin\vartheta\,\cos\psi)$$
$$Y = \underline{x}\,(-\cos\vartheta\,\cos\varphi\,\sin\psi - \sin\varphi\,\cos\psi) + \tag{6}$$
$$+ \underline{y}\,(-\cos\vartheta\,\sin\varphi\,\sin\psi + \cos\varphi\,\cos\psi) + \underline{z}\,\sin\vartheta\,\sin\psi$$
$$Z = \underline{x}\,\sin\vartheta\,\cos\varphi + \underline{y}\,\sin\vartheta\,\sin\varphi + \underline{z}\,\cos\vartheta$$

Further if u_x, u_y and u_z be the components of the angular speed of the vehicle as to the vehicle rigid system $X\,Y\,Z$, we read from the Figs. 1, 2 and 3:

$$u_X = -\dot{\varphi}\,\sin\vartheta\,\cos\psi + \dot{\vartheta}\,\sin\psi$$
$$u_Y = \dot{\varphi}\,\sin\vartheta\,\sin\psi + \dot{\vartheta}\,\cos\psi \tag{7}$$
$$u_Z = \dot{\psi} + \dot{\varphi}\,\cos\vartheta$$

2. Pitch moment and yaw moment due to main rocket motor in arbitrary position

For convenience let us introduce a vehicle rigid system of reference $x\,y\,z$, of which:

$$x = -X, y = Y, z = -Z,$$

as indicated in Fig. 4.

Let P be the point of intersection of the central line of the main nozzle and the roll axis, and let us write: $\overrightarrow{OP} = r_0'$ (O = centre of gravity of rocket vehicle + still included fuel). The direction of the main thrust as to the system $x\,y\,z$ is defined by the angles α and β (see Fig. 5). Then α is the angle included by $z+$ and the central line of the main thrust tube and β the angle included by the projection of this central line on the plane $x\,y$ and $x+$. Then denoting the main thrust by $S^{(M)}$, we obtain as moment of the main thrust as to O:

69

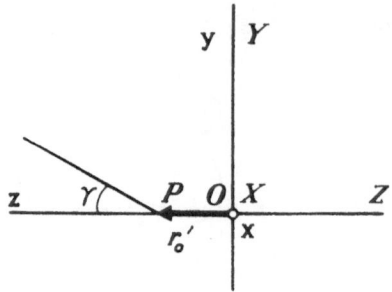

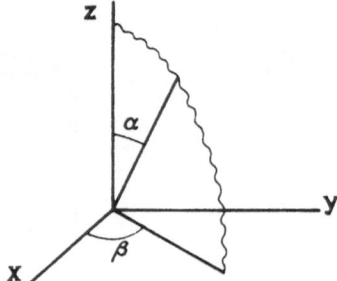

Fig. 4. $X+$ directed from observer.

Fig. 5.

$$\mathfrak{M}^{(M)} = r_0' \times S = \begin{vmatrix} \mathbf{i} & \mathbf{j} & \mathbf{k} \\ 0 & 0 & a \\ -S^{(M)} \sin\alpha \cos\beta & -S^{(M)} \sin\alpha \sin\beta & -S^{(M)} \cos\alpha \end{vmatrix}$$

in which $\mathbf{i} = \boldsymbol{i}_1, \mathbf{j} = \boldsymbol{j}_1, \mathbf{k} = -\boldsymbol{k}_1$ and $|\,r_0'\,| = a$.

Hence the components of the main thrust as to the system $X.Y Z$ become

$$\mathfrak{M}^{(M)}{}_x = -S^{(M)}a \sin\alpha \sin\beta, \quad \mathfrak{M}^{(M)}{}_y = -S^{(M)}a \sin\alpha \cos\beta, \quad \mathfrak{M}^{(M)}{}_z = 0 \qquad (8)$$

3. Roll moment due to the equal thrusts of the Vernier motors

Let us assume that the swivel mounted Vernier motors are arranged with their common swivel axis parallel to the $\mathbf{x}$ axis at distance a from the $\mathbf{x}$ axis in the $\mathbf{xy}$ plane. The points of intersection of the swivel axis and the nozzle axes be $\mathbf{z} = a$, $\mathbf{x} = x_0$ and $\mathbf{z} = a$, $\mathbf{x} = -x_0$. If S_V be the thrust of any Vernier motor and $\pm \gamma$ the deflection, we obtain as roll moment due to the Vernier motors:

$$\mathfrak{M}_z = 2S_V x_0 \sin\gamma. \qquad (9)$$

4. The gas flow in the nozzles

In order to account for the influence of the Coriolis accelerations of the fluid particles in the nozzles, when the vehicle rotates about an axis passing through the instantaneous centre of gravity of vehicle + still included fuel, we have to consider the gas flow in the thrust tubes. For simplicity let us assume one dimensional adiabatic flow.

We then have: [1]

$$G_{sec} = F_{min} \sqrt{2 \, \frac{\varkappa}{\varkappa + 1} \frac{p_0}{v_0} \left(\frac{2}{\varkappa + 1}\right)^{2/(\varkappa - 1)}} \qquad (10)$$

$$G_{sec} = F \sqrt{2 \, \frac{\varkappa}{\varkappa - 1} \frac{p_0}{v_0} \left[\left(\frac{p}{p_0}\right)^{2/\varkappa} - \left(\frac{p}{p_0}\right)^{(\varkappa + 1)/\varkappa}\right]} \qquad (11)$$

$$c = \sqrt{2 \, \frac{\varkappa}{\varkappa - 1} p_0 v_0 \left[1 - \left(\frac{p}{p_0}\right)^{(\varkappa - 1)/\varkappa}\right]} \qquad (12)$$

in which:

On the Dynamics of a Space Vehicle

G_{sec} = Mass quantum expelled per second
F_{min} = gorge cross section
F = any cross section
p_0 = pressure in combustion chamber
v_0 = inverse density of gas at entrance of nozzle
p = pressure in any cross section F
$\varkappa, \kappa$ = c_p/c_v, in which c_p = specific heat of gas flow at constant pressure
  and c_v = "   " " " " " " volume.

Let us further assume that the graph:

$$G_{sec} = G_{sec}(p_0) \tag{13}$$

be determined by experience. Then we can find the thrust S as function of p_0 in the following way. Assume for G_{sec} some value. Then by (13), p_0 is known and p_0/v_0 follows from (10), so that also v_0 becomes known. Further from (11) we find p/p_0 and consequently c from (12).
Then:

$$S = G_{sec}\ c_e$$

in which c_e = flow speed in mouth section. Hence in this way we find the graph:

$$S = S(G_{sec}) = S(p_0) \tag{14}$$

5. The Coriolis moment of a rocket motor with nozzle axis parallel to the z axis, but for the rest arbitrary position

The *Coriolis moment* is the steering moment required in order to give the fluid particles the corresponding Coriolis accelerations. Let $\mathbf{u}$ be the angular speed of the vehicle about the instantaneous axis of rotation passing through the instantaneous centre of gravity of vehicle + still included fuel. Further let $\mathbf{r}$ be the radius vector of the fluid particle ϱdV, as to the system $\mathbf{x\,y\,z}$, ϱ being the local fluid density and dV the volume of the particle. Then if $\mathbf{c}$ denotes the local speed of the fluid, we obtain as Coriolis moment:

$$\mathfrak{M}_{cor} = 2 \int [\mathbf{r} \times (\mathbf{u} \times \mathbf{c})]\, \varrho dV \tag{15}$$

in which the space integral must be extended throughout the fluid in the nozzle. Now if $\mathbf{r}_0$ = radius vector of centre of entrance section of nozzle, and $\mathbf{r} = \mathbf{r}_0 + \mathbf{r}'$, we obtain from (15):

$$\mathfrak{M}_{cor} = \mathbf{r}_0 \times 2 \int (\mathbf{u} \times \mathbf{c})\, \varrho dV + 2 \int [\mathbf{r}' \times (\mathbf{u} \times \mathbf{c})] \varrho dV \tag{16}$$

Now assuming one-dimensional flow, we have:

$$\mathbf{u} \times \mathbf{c} = \begin{vmatrix} \mathbf{i} & \mathbf{j} & \mathbf{k} \\ u_x & u_y & u_z \\ 0 & 0 & c(\mathbf{z}) \end{vmatrix} = \mathbf{i}u_y\, c(\mathbf{z}) - \mathbf{j}u_x\, c(\mathbf{z}).$$

Then, for convenience introducing a reference system $\mathbf{x'\,y'\,z'}$ parallel with $\mathbf{x\,y\,z}$, and with the centre of the entrance as origin, we obtain:

$$\mathbf{A} = \int (\mathbf{u} \times \mathbf{c})\, \varrho dV = \mathbf{i} u_y \int_0^l c(\mathbf{z}')\, \varrho(\mathbf{z}')\, F(\mathbf{z}')\, d\mathbf{z}' - \mathbf{i} u_x \int_0^l c(\mathbf{z}')\, \varrho(\mathbf{z}')\, F(\mathbf{z}')\, d\mathbf{z}' \tag{17}$$

in which $F(\mathbf{z}')$ denotes the cross section corresponding with $\mathbf{z}'$ and l denotes the length of the nozzle, taken along the central line.

Now:

$$p(\mathbf{z}')/p_0 = [\varrho(\mathbf{z}')/\varrho_0]^\kappa$$

so that:

$$\varrho(\mathbf{z}') = \varrho_0\, [p(\mathbf{z}')/p_0]^{1/\kappa} = (1/v_0)\, [p(\mathbf{z}')/p_0]^{1/\kappa}$$

Hence:

$$\int_0^l c(\mathbf{z}')\, \varrho(\mathbf{z}')\, F(\mathbf{z}')\, d\mathbf{z}' =$$

$$= \frac{1}{v_0} \int_0^l \sqrt{2\, \frac{\varkappa}{\varkappa - 1}\, p_0 v_0 \left[1 - \left(\frac{p(\mathbf{z}')}{p_0} \right)^{(\kappa - 1)/\kappa} \right]} \left(\frac{p(\mathbf{z}')}{p_0} \right)^{1/\kappa} F(\mathbf{z}')\, d\mathbf{z}' \tag{18}$$

By (17):

$$2\, \mathbf{r}_0 \times \int (\mathbf{u} \times \mathbf{c})\, \varrho dV = 2 \begin{vmatrix} \mathbf{i} & \mathbf{j} & \mathbf{k} \\ x_0 & y_0 & z_0 \\ A_x & A_y & 0 \end{vmatrix} \tag{19}$$

Further:

$$2\{\mathbf{r}' \times (\mathbf{u} \times \mathbf{c})\} = 2 \begin{vmatrix} \mathbf{i} & \mathbf{j} & \mathbf{k} \\ x' & y' & z' \\ u_y\, c(\mathbf{z}') & -u_x\, c(\mathbf{z}') & 0 \end{vmatrix} \tag{20}$$

Hence:

$$2 \int [\mathbf{r}' \times (\mathbf{u} \times \mathbf{c})]\, \varrho dV =$$

$$= 2\mathbf{i} u_x \int_0^l z'\, c(\mathbf{z}')\, \varrho(\mathbf{z}')\, F(\mathbf{z}')\, d\mathbf{z}' + 2\mathbf{j} u_y \int_0^l z'\, c(\mathbf{z}')\, \varrho(\mathbf{z}')\, F(\mathbf{z}')\, d\mathbf{z}' + \tag{21}$$

$$+ 2\mathbf{k}(-u_x \int x'\, c(\mathbf{z}')\, \varrho(\mathbf{z}')\, dx'\, dy'dz' - u_y \int y'\, c(\mathbf{z}')\, \varrho(\mathbf{z}')\, dx'\, dy'\, dz')$$

in which:

$$\int_0^l z'\, c(\mathbf{z}')\, \varrho(\mathbf{z}')\, F(\mathbf{z}')\, d\mathbf{z}' =$$

$$= \frac{1}{v_0} \int_0^l \sqrt{2\, \frac{\varkappa}{\varkappa - 1}\, p_0 v_0 \left[1 - \left(\frac{p(\mathbf{z}')}{p_0} \right)^{(\kappa - 1)/\kappa} \right]} \left(\frac{p(\mathbf{z}')}{p_0} \right)^{1/\kappa} z'\, F(\mathbf{z}')\, d\mathbf{z}' \tag{22}$$

Further we may write:

$$\iiint x'c(\mathbf{z}')\, \varrho(\mathbf{z}')\, dx'\, dy'\, dz' = \int c(\mathbf{z}')\, \varrho(\mathbf{z}')\, d\mathbf{z}' \int dy' \int x'\, dx'$$

The integration limits with respect to x' correspond with arbitrary values of y' and z'. We shall assume that the nozzle has circular cross sections. Let us indicate

On the Dynamics of a Space Vehicle

the radius of cross section $F(z')$ by $R(z')$. Then we obtain as equation of the circumference of this section:

$$x'^2 + y'^2 = [R(z')]^2.$$

The integration limits with respect to x' then become:

$$x'_1 = + \sqrt{[R(z')]^2 - y'^2}, \qquad x'_2 = - \sqrt{[R(z')]^2 - y'^2}$$

We then obtain:

$$\int_{x'_1}^{x'_2} x'\, dx' = [\tfrac{1}{2} x'^2]_{x'_1}^{x'_2} = 0 \tag{23}$$

Hence from (23) it follows that:

$$\int x'\, c(z')\, \varrho(z')\, dx'\, dy'\, dz' = 0, \quad \int y'\, c(z')\, \varrho(z')\, dx'\, dy'\, dz' = 0 \tag{24}$$

Then from (16), (17), (19), (21) and (24) it follows:

$$\mathfrak{M}_{cor_x} = 2z_0 u_x \int_0^l c(z')\, \varrho(z')\, F(z')\, dz' + 2u_x \int_0^l z'\, c(z')\, \varrho(z')\, F(z')\, dz'$$

$$\mathfrak{M}_{cor_y} = 2z_0 u_y \int_0^l c(z')\, \varrho(z')\, F(z')\, dz' + 2u_y \int_0^l z'\, c(z')\, \varrho(z')\, F(z')\, dz' \tag{25}$$

$$\mathfrak{M}_{cor_z} = -2x_0 u_z \int_0^l c(z')\, \varrho(z')\, F(z')\, dz' - 2y_0 u_z \int_0^l c(z')\, \varrho(z')\, F(z')\, dz'$$

Further it follows from (7) and Fig. 4 that:

$$\begin{aligned}
u_x &= -u_x = \dot{\varphi}\sin\vartheta\cos\psi - \dot{\vartheta}\sin\psi, \\
u_y &= u_y = \dot{\varphi}\sin\vartheta\sin\psi + \dot{\vartheta}\cos\psi, \\
u_z &= -u_z = -\dot{\psi} - \dot{\varphi}\cos\vartheta.
\end{aligned} \tag{4'}$$

6. Coriolis moment of main rocket motor in arbitrary position α, β

Assume firstly $\alpha = 0$, $\beta = 0$, so that the main rocket motor is in mid-position. In that case we have to put in the equations (25), $x_0 = 0$, $y_0 = 0$, $r_0 = kr_0$, so that we then obtain:

$$\mathfrak{M}_{cor_x} \Big|_{\alpha=0,\,\beta=0} = 2z_0 u_x \int_0^l c(z')\, \varrho(z')\, F(z')\, dz' + 2u_x \int_0^l z'\, c(z')\, \varrho(z')\, F(z')\, dz'$$

$$\mathfrak{M}_{cor_y} \Big|_{\alpha=0,\,\beta=0} = 2z_0 u_y \int_0^l c(z')\, \varrho(z')\, F(z')\, dz' + 2u_y \int_0^l z'\, c(z')\, \varrho(z')\, F(z')\, dz' \tag{25'}$$

$$\mathfrak{M}_{cor_z} \Big|_{\alpha=0,\,\beta=0} = 0$$

Now assume a new vehicle rigid system of reference $\underline{x}\,\underline{y}\,\underline{z}$, which is obtained from $x\,y\,z$ by a rotation β about $\underline{z}$ and a subsequent rotation α about $\underline{y}$. Then from Fig. 6 and Fig. 7 we read:

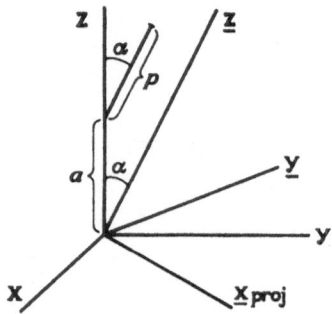

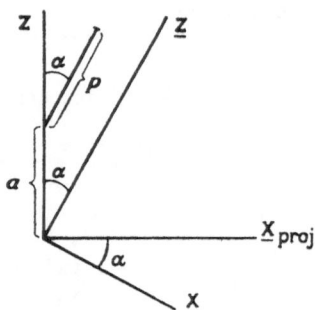

Fig. 6.   Fig. 7.

$$\cos(\underline{x}x) = \cos\alpha\,\cos\beta \qquad \cos(\underline{y}x) = -\sin\beta \qquad \cos(\underline{z}x) = \sin\alpha\,\cos\beta$$
$$\cos(\underline{x}y) = \cos\alpha\,\sin\beta \qquad \cos(\underline{y}y) = \cos\beta \qquad \cos(\underline{z}y) = \sin\alpha\,\sin\beta$$
$$\cos(\underline{x}z) = -\sin\alpha \qquad \cos(\underline{y}z) = 0 \qquad \cos(\underline{z}z) = \cos\alpha$$

so that:

$$\underline{x} = x\cos\alpha\,\cos\beta + y\cos\alpha\,\sin\beta - z\sin\alpha$$
$$\underline{y} = -x\sin\beta + y\cos\beta \tag{26}$$
$$\underline{z} = x\sin\alpha\,\cos\beta + y\sin\alpha\,\sin\beta + z\cos\alpha$$

Now if a denotes the distance of the rotation point P of the main rocket motor (situated on the $\underline{z}$ axis) to the origin of the system $\underline{x}\,\underline{y}\,\underline{z}$ ($=$ centre of gravity of vehicle $+$ still included fuel) and p denotes the distance from P to the centre of the entrance section of the main nozzle, we obtain as radius vector of this section centre:

$$\mathbf{r}_0 = \underline{\mathbf{i}}\underline{x}_0 + \underline{\mathbf{j}}\underline{y}_0 + \underline{\mathbf{k}}\underline{z}_0$$

$\underline{\mathbf{i}}$, $\underline{\mathbf{j}}$ and $\underline{\mathbf{k}}$ being unit vectors in $\underline{x}+$, $\underline{y}+$ and $\underline{z}+$ direction, whereas according to Figs. 6 and 7:

$$\underline{x}_0 = -a\sin\alpha,\, \underline{y}_0 = 0,\, \underline{z}_0 = p + a\cos\alpha$$

Now in the positions α, β the central axis of the main nozzle is parallel with the $\underline{z}$ axis, so that this position is analoguous to the position as to the system $\underline{x}\,\underline{y}\,\underline{z}$, as generally assumed in the equations (25).

$$\mathfrak{M}^{(\mathrm{M})}{}_{cor\underline{x}} = 2(p + a\cos\alpha)\, u_{\underline{x}}\int_0^l c(z')\, \varrho(z')\, F(z')\, dz' +$$
$$+ 2u_{\underline{x}}\int_0^l z'\, c(z')\, \varrho(z')\, F(z')\, dz'$$

$$\mathfrak{M}^{(\mathrm{M})}{}_{cor\underline{y}} = 2(p + a\cos\alpha)\, u_{\underline{y}}\int_0^l c(z')\, \varrho(z')\, F(z')\, dz' + \tag{27}$$
$$+ 2u_{\underline{y}}\int_0^l z'\, c(z')\, \varrho(z')\, F(z')\, dz'$$

$$\mathfrak{M}^{(\mathrm{M})}{}_{cor\underline{z}} = 2a\sin\alpha \cdot u_{\underline{x}}\int_0^l c(z')\, \varrho(z')\, F(z')\, dz'$$

On the Dynamics of a Space Vehicle

Further we have in (27), in connection with (26) and (4′):

$$u_{\underline{x}} = (\dot{\varphi} \sin\vartheta \cos\psi - \dot{\vartheta} \sin\psi) \cos\alpha \cos\beta + (\dot{\varphi} \sin\vartheta \sin\psi + \dot{\vartheta} \cos\psi) \cos\alpha \sin\beta +$$
$$+ (\dot{\psi} + \dot{\varphi} \cos\vartheta) \sin\alpha$$

$$u_{\underline{y}} = (- \dot{\varphi} \sin\vartheta \cos\psi + \dot{\vartheta} \sin\psi) \sin\beta + (\dot{\varphi} \sin\vartheta \sin\psi + \dot{\vartheta} \cos\psi) \cos\beta$$

$$u_{\underline{z}} = (\dot{\varphi} \sin\vartheta \cos\psi - \dot{\vartheta} \sin\psi) \sin\alpha \cos\beta + (\dot{\varphi} \sin\vartheta \sin\psi + \dot{\vartheta} \cos\psi) \sin\alpha \sin\beta +$$
$$- (\dot{\psi} + \dot{\varphi} \cos\vartheta) \cos\alpha \tag{28}$$

Further by (26) we have:

$$\mathfrak{M}^{(M)}{}_{cor_{\underline{x}}} = \mathfrak{M}^{(M)}{}_{cor_{x}} \cos\alpha \cos\beta - \mathfrak{M}^{(M)}{}_{cor_{y}} \sin\beta + \mathfrak{M}^{(M)}{}_{cor_{z}} \sin\alpha \cos\beta$$

$$\mathfrak{M}^{(M)}{}_{cor_{y}} = \mathfrak{M}^{(M)}{}_{cor_{x}} \cos\alpha \sin\beta + \mathfrak{M}^{(M)}{}_{cor_{y}} \cos\beta + \mathfrak{M}^{(M)}{}_{cor_{z}} \sin\alpha \sin\beta \tag{29}$$

$$\mathfrak{M}^{(M)}{}_{cor_{z}} = - \mathfrak{M}^{(M)}{}_{cor_{x}} \sin\alpha + \mathfrak{M}^{(M)}{}_{cor_{z}} \cos\alpha$$

By substitution of (27) in combination with (28) in (29), we then obtain $\mathfrak{M}^{(M)}{}_{cor_{x}}$, $\mathfrak{M}^{(M)}{}_{cor_{y}}$ and $\mathfrak{M}^{(M)}{}_{cor_{z}}$ as functions of $\dot{\vartheta}, \dot{\varphi}, \dot{\psi}, \vartheta, \psi, \alpha$ and β.

Further by Fig. 4:

$$\mathfrak{M}^{(M)}{}_{cor_{\underline{x}}} = - \mathfrak{M}^{(M)}{}_{cor_{x}}, \quad \mathfrak{M}^{(M)}{}_{cor_{\underline{y}}} = \mathfrak{M}^{(M)}{}_{cor_{y}}, \quad \mathfrak{M}^{(M)}{}_{cor_{\underline{z}}} = - \mathfrak{M}^{(M)}{}_{cor_{z}} \tag{30}$$

Hence in connection with (30) we ultimately find:

$$\mathfrak{M}^{(M)}{}_{cor_{\underline{x}}} = \mathfrak{M}^{(M)}{}_{cor_{\underline{x}}}(\dot{\vartheta}, \dot{\varphi}, \dot{\psi}, \vartheta, \psi, t)$$

$$\mathfrak{M}^{(M)}{}_{cor_{\underline{y}}} = \mathfrak{M}^{(M)}{}_{cor_{\underline{y}}}(\dot{\vartheta}, \dot{\varphi}, \dot{\psi}, \vartheta, \psi, t) \tag{31}$$

$$\mathfrak{M}^{(M)}{}_{cor_{\underline{z}}} = \mathfrak{M}^{(M)}{}_{cor_{\underline{z}}}(\dot{\vartheta}, \dot{\varphi}, \dot{\psi}, \vartheta, \psi, t)$$

The time coordinate t enters in the expressions (31), because by the fuel consumption $a = a(t)$, whereas also α and β are considered as prescribed functions of time as well as the integrals in (27) which again depend on S as function of time. Because the magnitudes of the thrusts are assumed to be prescribed as functions of t, also in connection with (14) the fuel consumption is prescribed as function of t, from which also a as function of time follows.

7. Coriolis moment of the Vernier motors

Introduce a vehicle rigid system of reference $x\ \underline{y}\ \underline{z}$, which is obtained from system $x\ y\ z$ by rotation γ about x in a sense as indicated in Fig. 8. (Hence then the rotation γ is in right handed relation with $X+$.) We then have transformation formulas:

and reversed:
$$\underline{y} = - z \sin\gamma + y \cos\gamma \qquad \underline{z} = z \cos\gamma + y \sin\gamma \tag{32}$$

$$y = \underline{z} \sin\gamma + \underline{y} \cos\gamma \qquad z = \underline{z} \cos\gamma - \underline{y} \sin\gamma \tag{33}$$

Let us indicate the Vernier motor corresponding with $+ x_{0} (= x_{0})$ by E and with $- x_{0}$ by F (compare Fig. 4). [Hence if γ is positive, the sense of the roll moment exerted by both Vernier motors is in right handed relation with $Z+$.]

Then the Vernier motor E, in a position γ, has its nozzle axis parallel with $\underline{z}+$, so that we have again a similar position as supposed in the equations (25) as to the system $x\ y\ z$. Hence in connection with (25) we obtain as components of the required Coriolis moment corresponding with the Vernier motor E:

$$\mathfrak{M}^{(E)}{}_{cor_x} = 2\underline{z}_0 u_x \int_0^{l_1}{}_{(E)}\underline{c}(z')\ \varrho(z')\ F(z')\ dz' + 2u_x \int_0^{l_1}{}_{(E)} z'\ c(z')\ \varrho(z')\ F(z')\ dz'$$

$$\mathfrak{M}^{(E)}{}_{cor_{\underline{y}}} = 2\underline{z}_0 u_{\underline{y}} \int_0^{l_1}{}_{(E)} c(z')\ \varrho(z')\ F(z')\ dz' + 2u_{\underline{y}} \int_0^{l_1}{}_{(E)} z'\ c(z')\ \varrho(z')\ F(z')\ dz' \quad (34)$$

$$\mathfrak{M}^{(E)}{}_{cor_{\underline{z}}} = -\ 2x_0 u_x \int_0^{l_1}{}_{(E)} c(z')\ \varrho(z')\ F(z')\ dz' - 2y_0 u_{\underline{y}} \int_0^{l_1}{}_{(E)} c(z')\ \varrho(z')\ F(z')\ dz'$$

in which l_1 denotes the length of the nozzle of any Vernier motor.

In (34) the integrals must now be extended along the central axis of the nozzle E. In order to indicate this, a lower suffix (E) is placed on the right of the integral signs. Further in (34), x_0, $\underline{y}_0$ and $\underline{z}_0$ now refer to the radiusvector $r_0{}^{(E)}$ of the entrance section of the nozzle E. Hence by Fig. 8:

$$r_0{}^{(E)} = \mathbf{i}x_0 + \underline{\mathbf{j}}y_0 + \underline{\mathbf{k}}z_0 = \mathbf{i}\,x_0 + \underline{\mathbf{j}}\,(-a\,\sin\gamma) + \underline{\mathbf{k}}\,(a\,\cos\gamma + q) \quad (35)$$

in which $\underline{\mathbf{j}}$ and $\underline{\mathbf{k}}$ denote unit vectors in the directions $\underline{y}+$ and $\underline{z}+$ and q denotes the distance from the centre of the entrance section of the nozzle of a Vernier motor to the swivel axis.

Further by (32):

$$u_{\underline{y}} = -\ u_z \sin\gamma + u_y \cos\gamma, \qquad u_{\underline{z}} = u_z \cos\gamma + u_y \sin\gamma.$$

Or in connection with (4'):

$$u_x = \dot{\varphi} \sin\vartheta \cos\psi - \dot{\vartheta} \sin\psi$$
$$u_{\underline{y}} = (\dot{\psi} + \dot{\varphi} \cos\vartheta) \sin\gamma + (\dot{\varphi} \sin\vartheta \sin\psi + \dot{\vartheta} \cos\psi) \cos\gamma \quad (36)$$
$$u_{\underline{z}} = -\ (\dot{\psi} + \dot{\varphi} \cos\vartheta) \cos\gamma + (\dot{\varphi} \sin\vartheta \sin\psi + \dot{\vartheta} \cos\psi) \sin\gamma$$

Hence substituting (36) in combination with (35) in (34) we obtain:

$$\mathfrak{M}^{(E)}{}_{cor_x} = 2(\dot{\varphi} \sin\vartheta \cos\psi - \dot{\vartheta} \sin\psi) \{(a \cos\gamma +$$
$$+\ q) \int_0^{l_1}{}_{(E)} c(z')\ \varrho(z')\ F(z')\ dz' + \int_0^{l_1}{}_{(E)} z'\ c(z')\ \varrho(z')\ F(z')\ dz'\}$$

$$\mathfrak{M}^{(E)}{}_{cor_{\underline{y}}} = 2\{(\dot{\psi} + \dot{\varphi} \cos\vartheta) \sin\gamma + (\dot{\varphi} \sin\vartheta \sin\psi + \dot{\vartheta} \cos\psi) \cos\gamma\} \times \quad (37)$$

$$\times \{(a \cos\gamma + q) \int_0^{l_1}{}_{(E)} c(z')\ \varrho(z')\ F(z')\ dz' + \int_0^{l_1}{}_{(E)} z'c(z')\ \varrho(z')\ F(z')\ dz'\}$$

$$\mathfrak{M}^{(\mathrm{E})}{}_{cor_{\underline{z}}} = -2x_0 \, (\dot{\varphi} \sin\vartheta \cos\psi - \dot{\vartheta} \sin\psi) \int_0^{l_1}{}_{(\mathrm{E})} c(\mathbf{z}') \, \varrho(\mathbf{z}') \, F(\mathbf{z}') \, d\mathbf{z}' +$$

$$+ \, 2[(\dot{\psi} + \dot{\varphi} \cos\vartheta) \sin\gamma + (\dot{\varphi} \sin\vartheta \sin\psi + \dot{\vartheta} \cos\psi) \cos\gamma] \times$$

$$\times \, a \sin\gamma \int_0^{l_1}{}_{(\mathrm{E})} c(\mathbf{z}') \varrho(\mathbf{z}') \, F(\mathbf{z}') \, d\mathbf{z}'$$

Further by (33) and Fig. 4:

$$\mathfrak{M}^{(\mathrm{E})}{}_{cor_{\mathbf{x}}} = - \; \mathfrak{M}^{(\mathrm{E})}{}_{cor_{\underline{x}}}$$

$$\mathfrak{M}^{(\mathrm{E})}{}_{cor_{\mathbf{y}}} = \; \mathfrak{M}^{(\mathrm{E})}{}_{cor_{\underline{z}}} \sin\gamma + \mathfrak{M}^{(\mathrm{E})}{}_{cor_{\underline{y}}} \cos\gamma = \mathfrak{M}^{(\mathrm{E})}{}_{cor_{Y}} \qquad (38)$$

$$\mathfrak{M}^{(\mathrm{E})}{}_{cor_{\mathbf{z}}} = \; \mathfrak{M}^{(\mathrm{E})}{}_{cor_{\underline{z}}} \cos\gamma - \mathfrak{M}^{(\mathrm{E})}{}_{cor_{\underline{y}}} \sin\gamma = - \; \mathfrak{M}^{(\mathrm{E})}{}_{cor_{Z}}$$

By (37) and (38) we then obtain:

$$\mathfrak{M}^{(\mathrm{E})}{}_{cor_{X}} = 2(-\dot{\varphi} \sin\vartheta \cos\psi + \dot{\vartheta} \sin\psi)\{(a \cos\gamma +$$

$$+ \, q) \int_0^{l_1}{}_{(\mathrm{E})} c(\mathbf{z}') \, \varrho(\mathbf{z}') \, F(\mathbf{z}') \, d\mathbf{z}' + \int_0^{l_1}{}_{(\mathrm{E})} \mathbf{z}' \, c(\mathbf{z}') \, \varrho(\mathbf{z}') \, F(\mathbf{z}') \, d\mathbf{z}'\}$$

$$\mathfrak{M}^{(\mathrm{E})}{}_{cor_{Y}} = 2\{-x_0 \, (\dot{\varphi} \sin\vartheta \cos\psi - \dot{\vartheta} \sin\psi) +$$

$$+ [(\dot{\psi} + \dot{\varphi} \cos\vartheta) \sin\gamma + (\dot{\varphi} \sin\vartheta \sin\psi + \dot{\vartheta} \cos\psi) \cos\gamma] \, a \sin\gamma\} \times$$

$$\times \, \sin\gamma \int_0^{l_1}{}_{(\mathrm{E})} c(\mathbf{z}') \, \varrho(\mathbf{z}') \; F(\mathbf{z}') \, d\mathbf{z}' +$$

$$+ \, 2\{(\dot{\psi} + \dot{\varphi} \cos\vartheta) \sin\gamma + (\dot{\varphi} \sin\vartheta \sin\psi + \dot{\vartheta} \cos\gamma)\} \times$$

$$\times \{(a \cos\gamma + q) \int_0^{l_1}{}_{(\mathrm{E})} c(\mathbf{z}') \, F(\mathbf{z}') \, d\mathbf{z}' + \int_0^{l_1}{}_{(\mathrm{E})} \mathbf{z}' \, c(\mathbf{z}') \, \varrho(\mathbf{z}') \, F(\mathbf{z}') \, d\mathbf{z}'\}$$

$$\mathfrak{M}^{(\mathrm{E})}{}_{cor_{Z}} = 2\{x_0 \, (\dot{\varphi} \sin\vartheta \cos\psi - \dot{\vartheta} \sin\psi) +$$

$$- [(\dot{\psi} + \dot{\varphi} \cos\vartheta) \sin\gamma + (\dot{\varphi} \sin\vartheta \sin\psi + \dot{\vartheta} \cos\psi) \cos\gamma] \, a_0 \sin\gamma\} \times$$

$$\times \, \cos\gamma \int_0^{l_1}{}_{(\mathrm{E})} c(\mathbf{z}') \, \varrho(\mathbf{z}') \, F(\mathbf{z}') \, d\mathbf{z}' +$$

$$+ \, 2\{(\dot{\psi} + \dot{\varphi} \cos\vartheta) \sin\gamma + (\dot{\varphi} \sin\vartheta \sin\psi + \dot{\vartheta} \cos\psi) \cos\gamma\} \sin\gamma \times$$

$$\times \{(a_0 \cos\gamma + q) \int_0^{l_1}{}_{(\mathrm{E})} c(\mathbf{z}') \, \varrho(\mathbf{z}') \, F(\mathbf{z}') \, d\mathbf{z}' + \int_0^{l_1}{}_{(\mathrm{E})} \mathbf{z}' \, c(\mathbf{z}') \, \varrho(\mathbf{z}') \, F(\mathbf{z}') \, d\mathbf{z}'\}$$

$$(39)$$

77

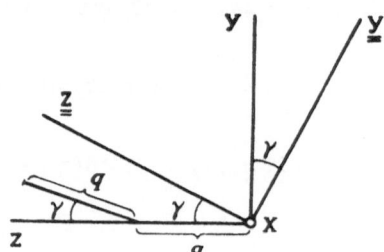

Fig. 8. x directed to observer.

We obtain the corresponding expressions for $\mathfrak{M}^{(F)}{}_{cor_X}$, $\mathfrak{M}^{(F)}{}_{cor_Y}$ and $\mathfrak{M}^{(F)}{}_{cor_Z}$ by replacing in (39) x_0 by $-x_0$ and γ by $-\gamma$. If we then write:

$$\mathfrak{M}^{(E)}{}_{cor_X} + \mathfrak{M}^{(F)}{}_{cor_X} = \mathfrak{M}^{\widehat{EF}}{}_{cor_X}$$
$$\mathfrak{M}^{(E)}{}_{cor_Y} + \mathfrak{M}^{(F)}{}_{cor_Y} = \mathfrak{M}^{\widehat{EF}}{}_{cor_Y}$$
$$\mathfrak{M}^{(E)}{}_{cor_Z} + \mathfrak{M}^{(F)}{}_{cor_Z} = \mathfrak{M}^{\widehat{EF}}{}_{cor_Z}$$

we obtain by adding:

$$\mathfrak{M}^{\widehat{EF}}{}_{cor_X} = 4(-\dot{\varphi}\sin\vartheta\cos\psi + \dot{\vartheta}\sin\psi)\{(a_0\cos\gamma +$$
$$+ q)\int_0^{l_1} {}_{(E)} c(z')\,\varrho(z')\,F(z')\,dz' + \int_0^{l_1} {}_{(E)} z'\,c(z')\,\varrho(z')\,F(z')\,dz'\}$$

$$\mathfrak{M}^{\widehat{EF}}{}_{cor_Y} = 4\{-x_0(\dot{\varphi}\sin\vartheta\cos\psi - \dot{\vartheta}\sin\psi) +$$
$$+ [(\dot{\psi} + \dot{\varphi}\cos\vartheta)\sin\gamma + (\dot{\varphi}\sin\vartheta\sin\psi + \dot{\vartheta}\cos\psi)\cos\gamma]\,a_0\sin\gamma\}\times$$
$$\times \sin\gamma \int_0^{l_1} {}_{(E)} c(z')\,\varrho(z')\,F(z')\,dz' +$$
$$+ 4(\dot{\varphi}\sin\vartheta\sin\psi + \dot{\vartheta}\cos\psi)\cos^2\gamma\{(a_0\cos\gamma + q)\times$$
$$\times \int_0^{l_1} {}_{(E)} c(z')\,\varrho(z')\,F(z')\,dz' + \int_0^{l_1} {}_{(E)} z'\,c(z')\,\varrho(z')\,F(z')\,dz'\}$$

$$\mathfrak{M}^{\widehat{EF}}{}_{cor_Z} = 4(\dot{\psi} + \dot{\varphi}\cos\vartheta)\sin^2\gamma\{(a_0\cos\gamma +$$
$$+ q)\int_0^{l_1} {}_{(E)} c(z')\,\varrho(z')\,F(z')\,dz' + \int_0^{l_1} {}_{(E)} z'\,c(z')\,\varrho(z')\,F(z')\,dz'\} \qquad (40)$$

8. Differential equations of rotary motion of the vehicle, in terms of ϑ, φ and ψ

Let us firstly assume that at any moment the vehicle rigid system $X\,Y\,Z$ coincides with the principal system of inertia of the vehicle + still included fuel. Then with respect to this system the vector differential equation of rotary motion becomes:

On the Dynamics of a Space Vehicle

$$\mathfrak{M} = \frac{d\mathbf{B}}{dt} + (\mathbf{u} \times \mathbf{B}) + (\mathbf{u} \times \boldsymbol{\omega} I) + \mathfrak{M}_{cor} \qquad (41)$$

In this equation $\mathfrak{M}$ denotes the resulting moment of the thrusts of the three rocket-motors with respect to O ($=$ centre of gravity of vehicle $+$ still included fuel); further $\mathbf{B}$ is the moment of momentum as to O of the vehicle as a whole, conceived as rigid body, with the rotor of the turbo pump assembly included, whereas $\boldsymbol{\omega}$ denotes the angular speed of this rotor as to the system $X\,Y\,Z$ and I its moment of inertia as to the axis of revolution. Let us assume that the direction of $\boldsymbol{\omega}$ coincides with the direction $X+$.

[*Remark:* If $\mathbf{v}_s$ denotes the tow speed of an arbitrary particle of the vehicle due to the rotation of the vehicle as a whole, and $\mathbf{v}'$ the relative speed of this particle as to the vehicle rigid system $X\,Y\,Z$, we may write for the total moment of momentum:

$$\mathbf{B}_{tot} = \Sigma m\,(\mathbf{r} \times \mathbf{v}) = \Sigma m\,(\mathbf{r} \times \mathbf{v}_s) + \Sigma m\,(\mathbf{r} \times \mathbf{v}'),$$

in which m denotes mass and $\mathbf{r}$ the radius vector of any particle and the summation be extended throughout all particles of the vehicle. Then:

$$\Sigma m\,(\mathbf{r} + \mathbf{v}_s) = \mathbf{B}.$$

Further:

$$\Sigma m\,(\mathbf{r} \times \mathbf{v}') = \Sigma m\{(\mathbf{r}_o + \mathbf{r}') \times \mathbf{v}'\} = \Sigma m\,(\mathbf{r}_o \times \mathbf{v}') + \Sigma m\,(\mathbf{r}' \times \mathbf{v}'),$$

in which $\mathbf{r}_o$ denotes the radius vector of the centre of gravity of the turbo pump assembly, whereas now the summation ought only to be extended throughout all particles of the rotor of the turbo pump assembly, because only for these particles $\mathbf{v}' \neq 0$. Then $\Sigma m\,(\mathbf{r}_o \times \mathbf{v}') = \mathbf{r}_o \times \Sigma m\mathbf{v}' = 0$ and $\Sigma(\mathbf{r} \times \mathbf{v}') = \mathbf{B}' = \boldsymbol{\omega}\,I$. Hence

$$\mathbf{B}_{tot} = \mathbf{B} + \boldsymbol{\omega}\,I,$$

so that:

$$\mathfrak{M}' = \left(\frac{d\mathbf{B}_{tot}}{dt}\right) = \frac{d\mathbf{B}}{dt} + (\mathbf{u} \times \mathbf{B}) + (\mathbf{u} \times \boldsymbol{\omega}\,I),$$

in which the second member denotes the rate of change of the moment of momentum of the vehicle $+$ still included fuel as to the system $\underline{x\,y\,z}$ which does not rotate as to the celestial background and $\mathfrak{M}' = \mathfrak{M} - \mathfrak{M}_{cor}$ denotes the total moment as to $\underline{x\,y\,z}$ exerted by the flowing gas on the vehicle $+$ still included fuel.]

Now in equation (41) we have:

$$\mathbf{u} \times \mathbf{B} = \begin{vmatrix} \mathbf{i}_1 & \mathbf{j}_1 & \mathbf{k}_1 \\ u_x & u_y & u_z \\ u_x I_x & u_y I_y & u_z I_z \end{vmatrix} \text{ and } \mathbf{u} \times \boldsymbol{\omega}\,I = \begin{vmatrix} \mathbf{i}_1 & \mathbf{i}_2 & \mathbf{i}_3 \\ u_x & u_y & u_z \\ \omega I & 0 & 0 \end{vmatrix}$$

in which I_X, I_Y and I_Z denote the moments of inertia of vehicle $+$ still included fuel as to X, Y and Z axis.

Hence the component equations of (41) become:

$$\mathfrak{M}_x = I_x \dot{u}_x + u_y u_z\,(I_z - I_y) + \mathfrak{M}_{cor_x}$$
$$\mathfrak{M}_y = I_y \dot{u}_y + u_x u_z\,(I_x - I_z) + u_z \omega I + \mathfrak{M}_{cor_y} \qquad (42)$$
$$\mathfrak{M}_z = I_z \dot{u}_z + u_x u_y\,(I_y - I_x) - u_y \omega I + \mathfrak{M}_{cor_z}$$

79

Further we find by differentiation of the equations (7):

$$\dot{u}_x = -\ddot{\varphi}\sin\vartheta\cos\psi - \dot{\varphi}\cos\vartheta\cos\psi\cdot\dot{\vartheta} + \dot{\varphi}\sin\vartheta\sin\psi\cdot\dot{\psi}$$

$$\dot{u}_y = \ddot{\varphi}\sin\vartheta\sin\psi + \dot{\varphi}\cos\vartheta\sin\psi\cdot\dot{\vartheta} + \dot{\varphi}\sin\vartheta\cos\psi\cdot\dot{\psi} + \ddot{\vartheta}\cos\psi - \dot{\vartheta}\sin\psi\cdot\dot{\psi}$$

$$\dot{u}_z = \ddot{\psi} + \ddot{\varphi}\cos\vartheta - \dot{\varphi}\sin\vartheta\cdot\dot{\vartheta} \tag{43}$$

Further by the equations (8) and (9) we obtain:

$$\mathfrak{M}_X = -S^{(M)}a_0\sin\alpha\sin\beta$$

$$\mathfrak{M}_Y = -S^{(M)}a_0\sin\alpha\cos\beta \tag{44}$$

$$\mathfrak{M}_Z = 2S_V x_0\sin\gamma$$

Further we obtain in (42):

$$\mathfrak{M}_{cor_x} = \mathfrak{M}^{(M)}_{cor_x} + \widehat{\mathfrak{M}^{EF}}_{cor_x}$$

$$\mathfrak{M}_{cor_y} = \mathfrak{M}^{(M)}_{cor_y} + \widehat{\mathfrak{M}^{EF}}_{cor_y} \tag{45}$$

$$\mathfrak{M}_{cor_z} = \mathfrak{M}^{(M)}_{cor_z} + \widehat{\mathfrak{M}^{EF}}_{cor_z}$$

Then substituting the equations (31) and the equations (40) in the right members of the equations (45) we obtain:

$$\mathfrak{M}_{cor_x} = \mathfrak{M}_{cor_x}(\dot{\varphi},\dot{\vartheta},\dot{\psi},\vartheta,\psi,t)$$

$$\mathfrak{M}_{cor_y} = \mathfrak{M}_{cor_y}(\dot{\varphi},\dot{\vartheta},\dot{\psi},\vartheta,\psi,t) \tag{46}$$

$$\mathfrak{M}_{cor_z} = \mathfrak{M}_{cor_z}(\dot{\varphi},\dot{\vartheta},\dot{\psi},\vartheta,\psi,t)$$

in which also γ is considered as prescribed function of time as well as the integrals in (40). Further also in (44) S, S_V, α and β are to be considered as prescribed functions of time, as well as I_x, I_y and I_z in (42) which also vary with the fuel consumption. (Thereby we assume that by the fuel consumption the system XYZ displaces parallel to itself as to the frame of the vehicle, whereas the Z axis still coincides with itself.)

Then substituting (7), (43), (44) and (46) in (42), we obtain three equations in

$$\vartheta,\psi,\dot{\vartheta},\dot{\varphi},\dot{\psi},\ddot{\vartheta},\ddot{\varphi},\ddot{\psi},\text{ and }t,$$

which are linear in the second derivatives $\ddot{\vartheta}$, $\ddot{\varphi}$ and $\ddot{\psi}$, and which can be solved as to those quantities. In this way we obtain the system:

$$\ddot{\vartheta} = f_1(\vartheta,\psi,\dot{\vartheta},\dot{\varphi},\dot{\psi},t),\quad \ddot{\varphi} = f_2(\vartheta,\psi,\dot{\vartheta},\dot{\varphi},\dot{\psi},t),\quad \ddot{\psi} = f_3(\vartheta,\psi,\dot{\vartheta},\dot{\varphi},\dot{\psi},t) \tag{47}$$

By writing $\dot{\vartheta} = \xi$, $\dot{\varphi} = \eta$ and $\dot{\psi} = \zeta$ the system (47) is equivalent with the system of 6 simultaneous differential equations of first order:

$$\dot{\xi} = f_1(\vartheta,\psi,\xi,\eta,\zeta,t),\dot{\eta} = f_2(\vartheta,\psi,\xi,\eta,\zeta,t),\dot{\zeta} = f_3(\vartheta,\psi,\xi,\eta,\zeta,t)$$

$$\dot{\vartheta} = \xi,\qquad \dot{\varphi} = \eta,\qquad \dot{\psi} = \zeta \tag{47'}$$

On the Dynamics of a Space Vehicle

The numerical integration can be carried out according to the method of Runge and Kutta, or by the method of successive approximations.

Applying the last one and starting from the initial state:

$$\vartheta_0, \varphi_0, \psi_0, \xi_0, \eta_0, \zeta_0 \text{ at time } t = 0,$$

we may write as solution in zero approximation:

$$\xi(t)_0 = \xi_0 + \dot{\xi}_0 t = \xi_0 + f_1(\vartheta_0, \psi_0, \xi_0, \eta_0, \zeta_0, t = 0)\, t$$

$$\eta(t)_0 = \eta_0 + \dot{\eta}_0 t = \eta_0 + f_2(\vartheta_0, \psi_0, \xi_0, \eta_0, \zeta_0, t = 0)\, t$$

$$\zeta(t)_0 = \zeta_0 + \dot{\zeta}_0 t = \zeta_0 + f_3(\vartheta_0, \psi_0, \xi_0, \eta_0, \zeta_0, t = 0)\, t$$

$$\vartheta(t)_0 = \vartheta_0 + \int_0^t (\xi_0 + \dot{\xi}_0 t)\, dt = \vartheta_0 + \xi_0 t + \tfrac{1}{2} f_{1_0} t^2 \qquad (48)$$

$$\varphi(t)_0 = \varphi_0 + \int_0^t (\eta_0 + \dot{\eta}_0 t)\, dt = \varphi_0 + \eta_0 t + \tfrac{1}{2} f_{2_0} t^2$$

$$\psi(t)_0 = \psi_0 + \int_0^t (\zeta_0 + \dot{\zeta}_0 t)\, dt = \psi_0 + \zeta_0 t + \tfrac{1}{2} f_{3_0} t^2$$

in which $f_{i_0} = f_i(\vartheta_0, \psi_0, \xi_0, \eta_0, \zeta_0, t = 0), i = 1, 2, 3.$

Substituting this zero approximation (48) in the equations (47') and integrating, we obtain as solution in first approximation:

$$\xi(t)_1 = \xi_0 + \int_0^t f_1[\vartheta(t)_0, \psi(t)_0, \xi(t)_0, \eta(t)_0, \zeta(t)_0, t]\, dt = \xi_0 + \chi_1(t)$$

$$\eta(t)_1 = \eta_0 + \int_0^t f_2[\vartheta(t)_0, \xi(t)_0, \eta(t)_0, \zeta(t)_0, t]\, dt = \eta_0 + \chi_2(t)$$

$$\zeta(t)_1 = \zeta_0 + \int_0^t f_3[\vartheta(t)_0, \psi(t)_0, \xi(t)_0, \eta(t)_0, \zeta(t)_0, t]\, dt = \zeta_0 + \chi_3(t)$$

$$\vartheta(t)_1 = \vartheta_0 + \int_0^t [\xi_0 + \chi_1(t)]\, dt = \vartheta_0 + \xi_0 t + \int_0^t \chi_1(t)\, dt \qquad (49)$$

$$\varphi(t)_1 = \varphi_0 + \int_0^t [\eta_0 + \chi_2(t)]\, dt = \varphi_0 + \eta_0 t + \int_0^t \chi_2(t)\, dt$$

$$\psi(t)_1 = \psi_0 + \int_0^t [\zeta_0 + \chi_3(t)]\, dt = \psi_0 + \zeta_0 t + \int_0^t \chi_3(t)\, dt$$

All integrals must be computed by quadratures. Then the first approximation (49) can again be substituted in the right members of the equations (47'), etc. In order to obtain a sufficient convergence, the time interval $0 \to t$ may not be taken too large. Then for the time t the initial conditions can be found with any required degree of accuracy, after which for a sufficiently small interval the same procedure of successive approximations can be applied, *etc.*

In case the principle system of inertia which we have assumed to be identical with $X\,Y\,Z$, does not displace parallel to itself as to the vehicle frame with the Z-axis

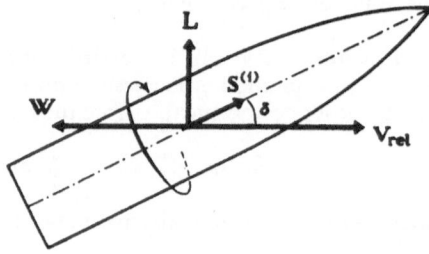

Fig. 9.

still coinciding with itself, we have to choose as system $X\,Y\,Z$ a system of reference, which displaces in that way. Thereby we still assume that the centre of gravity of rocket + still included fuel still remains on the roll axis, which is chosen as Z-axis.

In that more general case we have to write in vector equation (41):

$$\mathbf{u} \times \mathbf{B} = \begin{vmatrix} \mathbf{i}_1 & \mathbf{i}_2 & \mathbf{i}_3 \\ u_X & u_Y & u_Z \\ B_X & B_Y & B_Z \end{vmatrix}$$

in which:

$$B_X = u_X I_X - u_Y \Phi_{XY} - u_Z \Phi_{XZ}$$
$$B_Y = u_Y I_Y - u_X \Phi_{YX} - u_Z \Phi_{YZ}$$
$$B_Z = u_Z I_Z - u_X \Phi_{ZX} - u_Y \Phi_{ZY}$$

in which the symbols Φ_{ik} denote the corresponding centrifugal moments.

We then obtain instead of the equations (42):

$$\begin{aligned}
\mathfrak{M}_X &= I_X \dot{u}_X + u_Y \left(u_Z I_Z - u_X \Phi_{ZX} - u_Y \Phi_{ZY} \right) + \\
&\quad - u_Z \left(u_Y I_Y - u_X \Phi_{YX} - u_Z \Phi_{YZ} \right) + \mathfrak{M}_{\mathrm{cor}_X} \\
\mathfrak{M}_Y &= I_Y \dot{u}_Y + u_Z \left(u_X I_X - u_Y \Phi_{XY} - u_Z \Phi_{XZ} \right) + \\
&\quad - u_X \left(u_Z I_Z - u_X \Phi_{ZX} - u_Y \Phi_{ZY} \right) + u_Z \omega I + \mathfrak{M}_{\mathrm{cor}_Y} \\
\mathfrak{M}_Z &= I_Z \dot{u}_Z + u_X \left(u_Y I_Y - u_X \Phi_{YX} - u_Z \Phi_{YZ} \right) + \\
&\quad - u_Y \left(u_X I_X - u_Y \Phi_{XY} - u_Z \Phi_{XZ} \right) - u_Y \omega I + \mathfrak{M}_{\mathrm{cor}_Z}
\end{aligned} \tag{42'}$$

If the fuel consumption as function of time is prescribed, we may consider in (43') also the centrifugal moments Φ_{ik} as known functions of time. Then by again substituting (7), (43), (44) and (46) in (42'), we again obtain a system (47).

9. The differential equations of orbital motion of the centre of gravity of the vehicle

The motion of the vehicle is considered with respect to the geocentric system of reference $x\,y\,z$, which does not rotate with respect to the celestial sky. M be the mass of the vehicle + still included fuel at time $t = 0$. Then denoting the components of resulting thrust of the three rocket motors by S_x, S_y and S_z and the gravitational inertial field of force acting on the unit of mass of the vehicle by F_x, F_y and F_z, we obtain as differential equations of orbital motion:

$$\ddot{x} = \frac{S_x}{M - \int\limits_0^t G_{\text{sec}}(t)\, dt} + F_x(x, y, z, t),$$

$$\ddot{y} = \frac{S_y}{M - \int\limits_0^t G_{\text{sec}}(t)\, dt} + F_y(x, y, z, t), \tag{50}$$

$$\ddot{z} = \frac{S_z}{M - \int\limits_0^t G_{\text{sec}}(t)\, dt} + F_z(x, y, z, t)$$

(In the general case, F_x, F_y and F_z are also functions of t when the field of force also depends on the motions of celestial bodies as the Sun and the Moon, which may be considered as given functions of time.) Now we have by the previous:

$$S^{(M)}{}_X = S^{(M)} \sin\alpha \cos\beta, \quad S^{(M)}{}_Y = -S^{(M)} \sin\alpha \sin\beta, \quad S^{(M)}{}_Z = S^{(M)} \cos\alpha$$

$$\Sigma S_{v_X} = 0 \qquad\qquad \Sigma S_{v_Y} = 0 \qquad\qquad \Sigma S_{v_Z} = 2 S_V \cos\gamma,$$

so that: $\tag{51}$

$$S_X = S^{(M)} \sin\alpha \cos\beta, \quad S_Y = -S^{(M)} \sin\alpha \sin\beta, \quad S_Z = S^{(M)} \cos\alpha + 2 S_Y \cos\gamma$$

Further it follows from the equations (6), because the system $x\,y\,z$ is parallel with system $x\,y\,z$, that:

$$S_x = S_X (\cos\vartheta \cos\varphi \cos\psi - \sin\varphi \sin\psi) + S_Y (-\cos\vartheta \cos\varphi \sin\psi - \sin\varphi \cos\psi) +$$
$$+ S_Z (\sin\vartheta \cos\varphi)$$

$$S_y = S_X (\cos\vartheta \sin\varphi \cos\psi + \cos\varphi \sin\psi) + S_Y (-\cos\vartheta \sin\varphi \sin\psi + \cos\varphi \cos\psi) +$$
$$+ S_Z (\sin\vartheta \cos\varphi)$$

$$S_z = S_X (-\sin\vartheta \cos\psi) + S_Y (\sin\vartheta \sin\psi) + S_Z (\cos\vartheta) \tag{52}$$

In (51) and (52), $S^{(M)}$, S_V, α, β and γ are to be considered as prescribed functions of time. Then substituting (52) in combination with (51) in (50) we ultimately obtain:

$$\ddot{x} = \xi_1(x, y, z, t), \qquad \ddot{y} = \xi_2(x, y, z, t), \qquad \ddot{z} = \xi_3(x, y, z, t) \tag{53}$$

Then starting from an initial state x_0, y_0, z_0, $\dot{x}_0$, $\dot{y}_0$, $\dot{z}_0$ at time $t = 0$, the system (50) can again be integrated by steps, either by the method of Runge and Kutta, or by the method of successive approximations.

10. Flight through an atmosphere

In the foregoing, we have only considered the flight in world space. In case of flight through an atmosphere, the vector equation (41) becomes:

$$\mathfrak{M} = \frac{d\mathbf{B}}{dt} + (\mathbf{u} \times \mathbf{B}) + (\mathbf{u} \times \boldsymbol{\omega} I) + \mathfrak{M}_{\text{cor}} + \mathfrak{M}_{\text{aer}} \tag{54}$$

In Fig. 9 the curved arrow then indicates the positive sense of rotation of the aerodynamic moment $\mathfrak{M}_{aer}$, whereas $\mathbf{v}_{rel}$ denotes the relative speed of the vehicle as to the surrounding atmosphere. $\mathbf{S}^{(1)}$ denotes the unit vector along the roll axis.

Then

$$\angle(\mathbf{S}^{(1)}, \mathbf{v}_{rel}) = \delta$$

is the angle of incidence.

Let us assume that the shape of the vehicle is axially symmetric about the roll axis ($= Z$ axis). In that case we may write:

$$\mathfrak{M}_{aer} = \frac{\mathbf{S}^{(1)} \times \mathbf{v}_{rel}}{|\mathbf{S}^{(1)} \times \mathbf{v}_{rel}|} \, c_m(\delta) \, l \, \varrho(xyz) \, f(v_{rel})$$

in which

$$\delta = \arccos \frac{\mathbf{S}^{(1)} \mathbf{v}_{rel}}{v_{rel}} \text{ and } \mathbf{v}_{rel} = \mathbf{v} - (\mathbf{w} \times \mathbf{r})$$

(55)

where $\mathbf{v}$ denotes the speed of the vehicle as to the geocentric non rotating system of reference xyz, $c_m(\delta)$ a function of δ, l a certain length, $f(v_{rel})$ a function of v_{rel}, ϱ the local air density, and $\mathbf{w}$ the angular speed vector of terrestrial rotation, whereas $\mathbf{r}$ be the radius vector of the vehicle. The vector equation (54) again refers to the vehicle rigid system of reference $X Y Z$. From (55) it follows that:

$$\mathfrak{M}_{aer} = c_m(\delta) \, l \, \varrho(xyz) \, f(v_{rel})$$

(56)

If we take the geocentric non-rotating system xyz in such a way that the z axis coincides with the terrestrial axis of rotation, we obtain:

$$\mathbf{v}_{rel} = \mathbf{v} - (\mathbf{w} \times \mathbf{r}) = \mathbf{v} - \begin{vmatrix} \mathbf{i} & \mathbf{j} & \mathbf{k} \\ 0 & 0 & w \\ x & y & z \end{vmatrix}$$

(57)

Then we read from (57):

$$v_{rel_x} = \dot{x} + yw, \ v_{rel_y} = \dot{y} - xw, \ v_{rel_z} = \dot{z}$$

and

$$v_{rel} = \sqrt{(\dot{x} + yw)^2 + (\dot{y} - xw)^2 + \dot{z}^2}$$

(58)

Further:

$$\mathbf{S}^{(1)} = \mathbf{i} \sin\vartheta \cos\varphi + \mathbf{j} \sin\vartheta \sin\varphi + \mathbf{k} \cos\vartheta$$

so that:

$$\mathbf{S}^{(1)} \times \mathbf{v}_{rel} = \begin{vmatrix} \mathbf{i} & \mathbf{j} & \mathbf{k} \\ \sin\vartheta \cos\varphi & \sin\vartheta \sin\varphi & \cos\vartheta \\ \dot{x} + yw & \dot{y} - xw & \dot{z} \end{vmatrix} =$$

$$= \mathbf{i} \left\{ \sin\vartheta \sin\varphi \, \dot{z} - \cos\vartheta \cdot (\dot{y} - xw) \right\} + \mathbf{j} \left\{ (\dot{x} + yw) \cos\vartheta - \sin\vartheta \cos\varphi \cdot \dot{z} \right\} + \mathbf{k} \left\{ \sin\vartheta \cos\varphi \cdot (\dot{y} - xw) - \sin\vartheta \sin\varphi \cdot (\dot{x} + yw) \right\}$$

(59)

Hence in connection with (56) and (59) we obtain as component equations of (55):

On the Dynamics of a Space Vehicle

$$\mathfrak{M}_{aer_x} = \frac{[\sin\vartheta\sin\varphi \cdot \dot{z} - (\dot{y} - xw)\cos\vartheta]\, c_m(\delta)\, l\, \varrho(xyz)\, f(v_{rel})}{\sqrt{[\sin\vartheta\sin\varphi \cdot \dot{z} - (\dot{y} - xw)\cos\vartheta]^2 + [(\dot{x} + yw)\cos\vartheta - \sin\vartheta\cos\varphi \cdot \dot{z}]^2 +} \atop {+ [\sin\vartheta\cos\varphi \cdot (\dot{y} - xw) - \sin\vartheta\sin\varphi \cdot (\dot{x} + yw)]^2}}$$

$$\mathfrak{M}_{aer_y} = \frac{[(\dot{x} + yw)\cos\vartheta - \sin\vartheta\cos\varphi \cdot \dot{z}]\, c_m(\delta)\, l\, \varrho(xyz)\, f(v_{rel})}{\sqrt{[\sin\vartheta\sin\varphi \cdot \dot{z} - (\dot{y} - xw)\cos\vartheta]^2 + [(\dot{x} + yw)\cos\vartheta - \sin\vartheta\cos\varphi \cdot \dot{z}]^2 +} \atop {+ [\sin\vartheta\cos\varphi \cdot (\dot{y} - xw) - \sin\vartheta\sin\varphi \cdot (\dot{x} + yw)]^2}}$$

$$\mathfrak{M}_{aer_z} = \frac{[\sin\vartheta\cos\varphi \cdot (\dot{y} - xw) - \sin\vartheta\sin\varphi (\dot{x} + yw)]\, c_m(\delta)\, l\, \varrho(xyz)\, f(v_{rel})}{\sqrt{[\sin\vartheta\sin\varphi \cdot \dot{z} - (\dot{y} - xw)\cos\vartheta]^2 + [(\dot{x} + yw)\cos\vartheta - \sin\vartheta\cos\varphi \cdot \dot{z}]^2 +} \atop {+ [\sin\vartheta\cos\varphi \cdot (\dot{y} - xw) - \sin\vartheta\sin\varphi \cdot (\dot{x} + yw)]^2}}$$

$$\tag{60}$$

in which:

$$\delta = \arccos \frac{\sin\vartheta\cos\varphi \cdot (\dot{x} + yw) + \sin\vartheta\sin\varphi \cdot (\dot{y} - xw) + \cos\vartheta \cdot \dot{z}}{\sqrt{(\dot{x} + yw)^2 + (\dot{y} - xw)^2 + \dot{z}^2}}$$

Then further by the transformation equations (6) we obtain:

$$\mathfrak{M}_{aer_X} = \mathfrak{M}_{aer_x}\,(\cos\vartheta\cos\varphi\cos\psi - \sin\varphi\sin\psi) +$$
$$+ \mathfrak{M}_{aer_y}\,(\cos\vartheta\sin\varphi\cos\psi + \cos\varphi\sin\psi) + \mathfrak{M}_{aer_z}\,(-\sin\vartheta\cos\psi)$$

$$\mathfrak{M}_{aer_Y} = \mathfrak{M}_{aer_x}\,(-\cos\vartheta\cos\varphi\sin\psi - \sin\varphi\cos\psi) + \tag{61}$$
$$+ \mathfrak{M}_{aer_y}\,(-\cos\vartheta\sin\varphi\sin\psi + \cos\varphi\cos\psi) + \mathfrak{M}_{aer_z}\,\sin\vartheta\sin\psi$$

$$\mathfrak{M}_{aer_Z} = 0$$

(That $\mathfrak{M}_{aer_Z} = 0$ follows directly from the fact that $\mathfrak{M}_{aer}$ has the direction of $(\mathbf{S}^{(1)} \times \mathbf{v}_{rel})$, so that $\mathfrak{M}_{aer}$ is perpendicular to $\mathbf{S}^{(1)}$, that is perpendicular to the direction $Z+$.) By substituting (60) in (61) we then obtain:

$$\mathfrak{M}_{aer_X} = \mathfrak{M}_{aer_X}(x, y, z, \dot{x}, \dot{y}, \dot{z}, \vartheta, \varphi, \psi)$$
$$\mathfrak{M}_{aer_Y} = \mathfrak{M}_{aer_Y}(x, y, z, \dot{x}, \dot{y}, \dot{z}, \vartheta, \varphi, \psi) \tag{62}$$
$$\mathfrak{M}_{aer_Z} = 0$$

Further we obtain as component equations of (54):

$$\mathfrak{M}_X = I_X \dot{u}_X + u_Y u_Z\,(I_Z - I_Y) + \mathfrak{M}_{cor_X} + \mathfrak{M}_{aer_X}$$
$$\mathfrak{M}_Y = I_Y \dot{u}_Y + u_X u_Z\,(I_X - I_Z) + u_Z\omega I + \mathfrak{M}_{cor_Y} + \mathfrak{M}_{aer_Y} \tag{63}$$
$$\mathfrak{M}_Z = I_Z \dot{u}_Z + u_X u_Y\,(I_Y - I_X) - u_Y\omega I + \mathfrak{M}_{cor_Z}$$

in which again I_X, I_Y and I_Z may be considered as known functions of time, in connection with the fuel consumption.

Then again substituting (7), (43), (44), (46) and (62) in (63), we obtain three equations in ϑ, φ, ψ, $\dot{\vartheta}$, $\dot{\varphi}$, $\dot{\psi}$, $\ddot{\vartheta}$, $\ddot{\varphi}$, $\ddot{\psi}$, x, y, z, $\dot{x}$, $\dot{y}$, $\dot{z}$, t, which are linear as to $\ddot{\vartheta}$,

φ, and ψ. Then solving the so obtained equations with respect to $\ddot{\vartheta}$, $\ddot{\varphi}$, and $\ddot{\psi}$, we ultimately obtain:

$$\ddot{\vartheta} = f_1(\vartheta, \varphi, \psi, \dot{\vartheta}, \dot{\varphi}, \dot{\psi}, x, y, z, \dot{x}, \dot{y}, \dot{z}, t)$$

$$\ddot{\varphi} = f_2(\vartheta, \varphi, \psi, \dot{\vartheta}, \dot{\varphi}, \dot{\psi}, x, y, z, \dot{x}, \dot{y}, \dot{z}, t) \tag{64}$$

$$\ddot{\psi} = f_3(\vartheta, \varphi, \psi, \dot{\vartheta}, \dot{\varphi}, \dot{\psi}, x, y, z, \dot{x}, \dot{y}, \dot{z}, t)$$

From (64) it follows that, as soon as aerodynamic moments come into play, the orbital motion of the centre of gravity of rocket + still included fuel becomes interdependent with the rotary motion about the instantaneous centre of gravity of vehicle + still included fuel. Now in this case of flight through the atmosphere the vector differential equation of the orbital motion becomes:

$$\ddot{\mathbf{r}} = \frac{\mathbf{S} + \mathbf{L} + \mathbf{W}}{M - \int\limits_0^t G_{\mathrm{sec}}(t)\,dt} + \mathbf{F}(x, y, z, t) \tag{65}$$

in which $\mathbf{F}(x, y, z, t)$ again indicates gravitational inertial field of force referring to the unit of mass by which the vehicle is acted upon, whereas $\mathbf{L}$ denotes the lift and $\mathbf{W}$ the drag. The direction of $\mathbf{W}$ then coincides with $-\mathbf{v}_{\mathrm{rel}}$, whereas the direction of $\mathbf{L}$ coincides with $\mathbf{v}_{\mathrm{rel}} \times (\mathbf{S}^{(1)} \times \mathbf{v}_{\mathrm{rel}})$, in which by the previous:

$$\mathbf{S}^{(1)} = \mathbf{i} \sin\vartheta \cos\varphi + \mathbf{j} \sin\vartheta \sin\varphi + \mathbf{k} \cos\vartheta$$

and

$$\mathbf{v}_{\mathrm{rel}} = \mathbf{i}\,(\dot{x} + yw) + \mathbf{j}\,(\dot{y} - xw) + \mathbf{k}\,\dot{z}$$

Then:

$$\mathbf{W} = -\frac{\mathbf{v}_{\mathrm{rel}}}{v_{\mathrm{rel}}}\, c_w(\delta)\, \varrho(xyz)\, F(v_{\mathrm{rel}}) \tag{66}$$

and

$$\mathbf{L} = \frac{\mathbf{v}_{\mathrm{rel}} \times (\mathbf{S}^{(1)} \times \mathbf{v}_{\mathrm{rel}})}{\left| \mathbf{v}_{\mathrm{rel}} \times (\mathbf{S}^{(1)} \times \mathbf{v}_{\mathrm{rel}}) \right|}\, c_a(\delta)\, \varrho(xyz)\, F(v_{\mathrm{rel}}) \tag{67}$$

in which again $\delta = \arccos (\mathbf{S}^{(1)}\mathbf{v}_{\mathrm{rel}}/v_{\mathrm{rel}})$. Then in (67):

$$\mathbf{v}_{\mathrm{rel}} \times (\mathbf{S}^{(1)} \times \mathbf{v}_{\mathrm{rel}}) = \mathbf{v}_{\mathrm{rel}} \times \begin{vmatrix} \mathbf{i} & \mathbf{j} & \mathbf{k} \\ \sin\vartheta \cos\varphi & \sin\vartheta \sin\varphi & \cos\vartheta \\ \dot{x} + yw & \dot{y} - xw & \dot{z} \end{vmatrix} =$$

$$= \begin{vmatrix} \mathbf{i} & \mathbf{j} & \mathbf{k} \\ \dot{x} + yw & \dot{y} - xw & \dot{z} \\ \dot{z}\sin\vartheta \sin\varphi + & (\dot{x} + yw)\cos\vartheta + & \sin\vartheta \cos\varphi\,(\dot{y} - xw) + \\ - (\dot{y} - xw)\cos\vartheta & -\dot{z}\sin\vartheta \cos\varphi & -(\dot{x} + yw)\sin\vartheta \sin\varphi \end{vmatrix} \tag{68}$$

Hence by (66), (67) and (68), the components W_x, W_y, W_z, L_x, L_y, and L_z are defined as functions of $x, y, z, \dot{x}, \dot{y}, \dot{z}, \varphi$ and ϑ.

On the Dynamics of a Space Vehicle

Let us write for simplicity:

$$W_x = W_x(x, y, z, \dot{x}, \dot{y}, \dot{z}, \varphi, \vartheta) \qquad L_x = L_x(x, y, z, \dot{x}, \dot{y}, \dot{z}, \varphi, \vartheta)$$

$$W_y = W_y(x, y, z, \dot{x}, \dot{y}, \dot{z}, \varphi, \vartheta) \qquad L_y = L_y(x, y, z, \dot{x}, \dot{y}, \dot{z}, \varphi, \vartheta) \qquad (69)$$

$$W_z = W_z(x, y, z, \dot{x}, \dot{y}, \dot{z}, \varphi, \vartheta) \qquad L_z = L_z(x, y, z, \dot{x}, \dot{y}, \dot{z}, \varphi, \vartheta)$$

Then by the equations (51) (in which $S^{(M)}$ and S_V are prescribed functions of t), (52) and (69) the component equations of (65) are defined, which then contain the quantities $x, y, z, \dot{x}, \dot{y}, \dot{z}, \vartheta, \varphi, \psi$ and t. Let us shortly denote these component equations of (65) by writing:

$$\ddot{x} = \zeta_1(x, y, z, \dot{x}, \dot{y}, \dot{z}, \vartheta, \varphi, \psi, t)$$

$$\ddot{y} = \zeta_2(x, y, z, \dot{x}, \dot{y}, \dot{z}, \vartheta, \varphi, \psi, t) \qquad (70)$$

$$\ddot{z} = \zeta_3(x, y, z, \dot{x}, \dot{y}, \dot{z}, \vartheta, \varphi, \psi, t)$$

Hence when aerodynamic forces and moments come into play, the equations (64) of the rotary motion about the centre of gravity of vehicle + still included fuel and the equations (70) of the orbital motion of this centre of gravity are interdependent, and can then only be solved in combination, starting from an initial state:

$$t = 0 \to x_0, y_0, z_0, \dot{x}_0, \dot{y}_0, \dot{z}_0, \vartheta_0, \varphi_0, \psi_0, \dot{\vartheta}_0, \dot{\varphi}, \dot{\psi}_0.$$

For the solution of this combined system (64), (70) the method of Runge and Kutta is most suitable.

Reference
[1] *Kooy* and *Uytenbogaart,* Ballistics of the Future, page 308, point 7 (Haarlem, The Technical Publishing Company, H. Stam, 1946).

Attitude Control of Rockets Using a Single Axis Control Jet

R. M. Howe

University of Michigan, Ann Arbor, Mich., USA

Complete attitude control of a rocket during both powered flight and coasting can be achieved with an auxiliary control-jet along a single body axis, providing the rocket is spun about its longitudinal axis at a modest rate. Then by timing the control-jet impulse to occur during the proper portion of each spin revolution, change of attitude angle in any direction can be achieved. Effect of main-engine thrust misalignment is mostly cancelled by the slow rocket spin, so that only modest impulses are needed from the control-jet to maintain attitude. The equations of motion describing this scheme are developed and electronic differential analyzer solutions which demonstrate system performance are presented. Transients are damped using a rate gyro or a linear accelerometer as a sensing device. Mechanization of the control method using an attitude reference gyro or a horizon scanner is described. Compressed air, hydrogen peroxide, and solid fuel hot gas sources for control-jet impulses are discussed. A proposed autopilot system with only one moving part, weighing approximately 5—10 lb, and capable of attitude controlling a 500—1000 lb rocket during and after powered flight in a vacuum appears feasible.

The success of simple uncontrolled sounding rockets such as the Aerobee and Nike-Cajun over the past years is well-known. These rockets have served as low-cost, reliable vehicles to carry instrumentation up to altitudes of 160 miles. More recently, similar uncontrolled vehicles employing more stages have been fired to altitudes of several hundred miles. Present sounding-rocket requirements indicate that altitudes of 1000 miles and higher are needed to perform desirable experiments. Because of the conflicting requirements of aerodynamic heating and aerodynamic stability, and because of the potentially large dispersion of multistage rockets capable of altitudes of 1000 miles and higher, it is generally believed that some form of attitude control is desirable and even necessary to control the trajectory of such rockets and allow the upper stage or stages to burn at altitudes beyond the region where aerodynamic stabilization is effective. For the space-probe of rocket, which may penetrate to distances of tens of thousands of miles, the requirement of attitude control is even more evident.

Then, too, many of the experiments which one would like to carry out with sounding rockets (*e.g.,* photogrammetric experiments) would benefit from attitude control during the free-flight portion of the trajectory. Also, good trajectory control may allow the recovery of certain experiments through reasonable control of point of impact.

As the state of the art in rocket engines advances, the dependence on aerodynamic stabilization of sounding rockets becomes more of a handicap and an attitude-control system becomes more necessary. This is obviously true for long-burning multistage liquid rockets. It is also true for the solid rockets such as ARCON and IRIS which are end-burners and have long burning times. Furthermore, new developments, such as spherical solid rockets, promise tremendous improvements in attainable mass ratios per stage, which can reduce either the number of stages required or the take-off weight for a given payload. However, the drag penalty of a spherical rocket almost precludes the use of aerodynamic stabilization because of the large cross-sectional area. The longer burning time also presents a problem, and to take any considerable advantage of such advances in solid rocketry, an attitude-control system appears to be necessary.

One further advantage of an attitude-control system it that it allows guidance to be employed with the rocket. This permits accurate launchings of satellites and other space vehicles.

Of course, attitude-control systems for both liquid and solid rockets have been in use for some time. It is felt that current systems employing such schemes as gimballed engines and jetevators are much too complex and expensive for sounding-rocket applications. For this reason the High Altitude Engineering Laboratory of the Department of Aeronautical and Astronautical Engineering at The University of Michigan has spent some time considering a number of possible techniques for achieving a simple attitude-control system for sounding rockets. The purpose of this paper is to describe a simple technique of attitude control and to discuss the physical implementation of the proposed scheme.

Description of the Spin-Modulation Method of Attitude Control

The main feature of this proposed attitude control method is that it requires only a single reaction-jet to control both pitch and yaw angle. Assume that a single auxiliary control-jet is located forward of the rocket c.g. (say, in the payload section) with its thrust axis perpendicular to the longitudinal axis of the rocket. Next assume that the rocket is spun about its longitudinal axis at a modest rate. By opening the control jet during that portion of a spin revolution when the nozzle is pointed in the correct direction, a pitching velocity impulse of correct sign and direction can be imparted. Thus attitude control with only one control-jet can be mechanized. Assuming that this scheme is only used when aerodynamic forces are negligible, the control-jet thrust level can be very modest, since most of the effect of any main-engine thrust misalignment is cancelled by the spin of the rocket. Furthermore, this attitude controller can be used during coasting periods as well as thrusting periods. Because it is potentially a very lightweight system, it has the further advantage of suggesting the possibility of the attitude control of several stages with a single controller in the payload section of the last stage.

Use of a Free Gyro for Directional Reference

Before considering a more detailed analysis of the spin-modulated attitude control technique described in the previous paragraph, let us consider several ways of obtaining directional reference and the signals needed to turn on and off the single-axis control-jet. The first method would employ a free gyro as a directional reference, with the spin-axis of the gyro aligned along the desired attitude direction. A hemispherical dome attached to the rocket case could be mounted directly over the gyro case. A light source (either active or a mirror) would be mounted on top of the gyro case directly on the spin axis. If the rocket is aligned in the correct direction (*i.e.,* with its longitudinal axis in the same direction as the gyro spin axis), the light on top of the gyro case will fall directly beneath the center of the hemispherical dome attached to the rocket. On the other hand, if the attitude angle of the rocket is not correct, the light will fall off to one side of the center of the dome. As the rocket spins, the light will describe a circular motion about the center of the dome, the circle radius being proportional to the attitude error. A pie-shaped transparent segment cut out of the dome will allow the light on the gyro case beneath the dome to be visible above the dome once each revolution of the rocket. If this transparent sector is aligned parallel to the direction of the control-jet, the jet can be turned on whenever a photocell sees the gyro light beneath the sector. In this way the necessary modulation of the control-jet is achieved, using a free gyro as the directional reference. By applying prescribed torques to the gyro about the horizontal axis, the gyro could be precessed through programmed attitude changes and the control system would make the rocket attitude change accordingly.

Consultation with gyro manufactures indicates that a small air-driven free gyro with air bearings is practical for the above operation. It is felt that accuracies of the order of one degree or better can be held for several minutes, even under intermittent acceleration fields up to $60\,g$. However, slower burning solid rockets and liquid rockets may limit the peak acceleration to values well below $60\,g$, in which case a free gyro should most certainly be an accurate attitude reference. It is felt that a gas-driven gyro is preferable to an electrically driven gyro because of the simplicity of carrying the small amount of gas needed.

Use of a Horizon Scanner for Vertical Reference

For vertical sounding rockets, a photoelectric horizon scanner appears very attractive. By having a rather broad look angle in elevation, the horizon can be seen over a wide range of altitude. When the spinning rocket is vertical, there will be no appreciable modulation of intensity in the signal received by the photocell. When the rocket attitude is off from the vertical, the photocell will, alternately, see a darker and lighter image, giving rise to a periodic modulation of intensity at the spin frequency of the rocket. After filtering, amplification, and differentiation, the signal will be a pulse which has one sign when the photocell is sweeping through the horizon from sky to earth and the opposite sign when the photocell is sweeping from earth to sky. By proper alignment of the photocell with respect to the control-jet, one of the above pulses can be used to turn on the jet for a short burst. This will occur when the nozzle has dipped the maximum amount below the horizon, and hence will apply an erecting impulse to the rocket. To accomplish this, it is evident that the nozzle should be located 90° around the rocket from the photocell scanner location. If two nozzles facing in opposite directions are used, the gas source can be turned on at all times and merely deflected through one nozzle or the other by a flapper valve. Under these conditions, a positive pulse from the photocell circuit can be used to divert the gas through the first nozzle, while a negative pulse sends the gas through the second nozzle.

Since the output signal from the photocell scanner is sent through a high-pass filter with a low-frequency cutoff somewhat below the rocket spin frequency, any steady component in the signal is removed. This makes the system insensitive to the average intensity about which the periodic modulation occurs. The filtered signal is then amplified to produce a square wave, which is subsequently differentiated to provide the control pulses. These pulses will be essentially independent of photocell sensitivity and of the amplitude of modulation. Details such as the best wavelengths to scan, the effect of the sun, possible use of infrared at night, *etc.*, still need to be considered. However, the horizon-scan method is very attractive because of its low weight, relative insensitivity to high acceleration, and its basic simplicity.

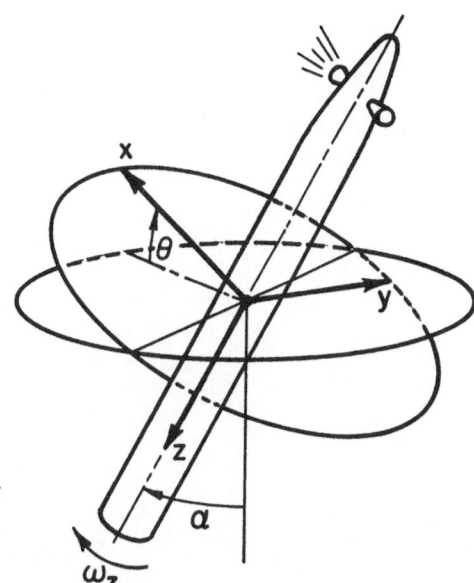

Fig. 1. Coordinate System for Representing
Rocket Orientation.

Method for Damping the Spin-Modulated Attitude Control System

To study the dynamics of the spin-modulation attitude control system, the rotational equations of motion of the rigid-body rocket were set up on the electronic differential analyzer. The usual orthogonal right-hand set of axes x, y, and z attached to the rocket with origin at the rocket c.g. were employed, except that the z-axis was assumed to be the axis of symmetry of the rocket, the x- and y-axes lying in the plane perpendicular to the longitudinal axis (see Fig. 1). The rocket angular velocity components ω_x, ω_y, and ω_z were solved for. Since no moments are applied about the z-axis, and since $I_x = I_y$ (the moments of inertia about the x- and y-axes are equal), it turns out that ω_z is a constant equal to the initial spin velocity of the rocket. The equations for ω_x and ω_y become simply

$$\dot{\omega}_x = (1 - I_z/I_x)\, \omega_z\omega_y + L_x$$

$$\dot{\omega}_y = - (1 - I_z/I_x)\, \omega_z\omega_x + L_y \tag{1}$$

where L_x and L_y are the external moments due to the jet control and where I_z/I_x is the ratio of the moment of inertia about the spin axis to that about the pitch axis. For long slender rockets $I_z/I_x \ll 1$. Since ω_z is a constant, these equations represent an undamped second-order linear system with natural frequency approximately ω_z, the rocket spin frequency. This concept is important in considering the proposed damping method to be discussed subsequently.

To determine the appropriate time to apply erecting impulses L_x or L_y, and to compute the rocket attitude angle Θ, the three direction cosines l_3, m_3, and n_3, representing projections of unit x-, y-, and z-vectors onto the vertical, were computed from ω_x, ω_y, and ω_z. Thus

$$\dot{l}_3 = m_3\omega_z - n_3\omega_y$$
$$\dot{m}_3 = n_3\omega_x - l_3\omega_z \tag{2}$$
$$\dot{n}_3 = l_3\omega_y - m_3\omega_x$$

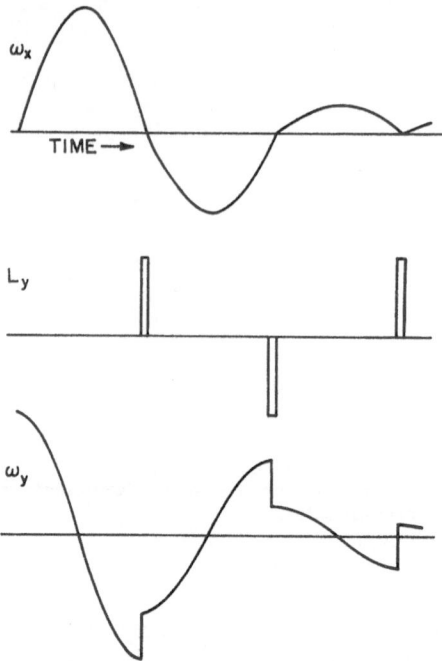

Fig. 2. Method of Damping ω_x and ω_y by
Torque Impulses.

The rocket attitude angle Θ can be obtained from $l_3 = -\sin\Theta$. Note in Fig. 1 that $\Theta = 0$ when the rocket is oriented vertically, and that, when the z-axis is off the vertical (as in the figure), Θ will oscillate symmetrically about zero as the rocket spins about its z-axis. m_3 is given by $\sin\Phi\cos\Theta$, so that when m_3 goes through zero, the bank angle Φ goes through zero. This occurs when the y-axis passes through the horizontal plane and hence at the time a horizon scanner would sweep the horizon. At this instant an erecting impulse torque L_y about the y-axis should be applied. Thus m_3 becomes the variable which, when it goes through zero, indicates the appropriate time to fire the control-jet.

One of the most important considerations in this type of control system is the method employed to damp the transients. One technique involves the measurement of $\dot{\alpha}$, the rate of change of the z-axis attitude angle. One might assume that, when $\dot{\alpha}$ and α have the same sign, a larger impulse should be applied; conversely, when $\dot{\alpha}$ and α have opposite signs, a smaller impulse should be applied. Our differential analyzer studies indicate that, unfortunately, this scheme will not work. The system eventually reaches a condition where the rocket motion is purely precessional.

The main function of damping should be to remove energy from the mode which is being controlled. In our case here this means the removal of energy associated with angular velocity components ω_x and ω_y in the plane perpendicular to the rocket spin axis. Reference to Equation 1 again shows that the dynamics of ω_x and ω_y correspond to an undamped second-order system, and that the erecting impulses L_y excite a transient in ω_x and ω_y which is approximately at the spin frequency ω_z. The function of the damping should be to reduce this undamped transient. Clearly this can be achieved by applying an impulse L_y of proper sign whenever ω_x goes through zero, as shown in Fig. 2. The instrumentation for this can be a rate gyro with its axis of sensitivity along the x rocket axis. By amplifying and differentiating

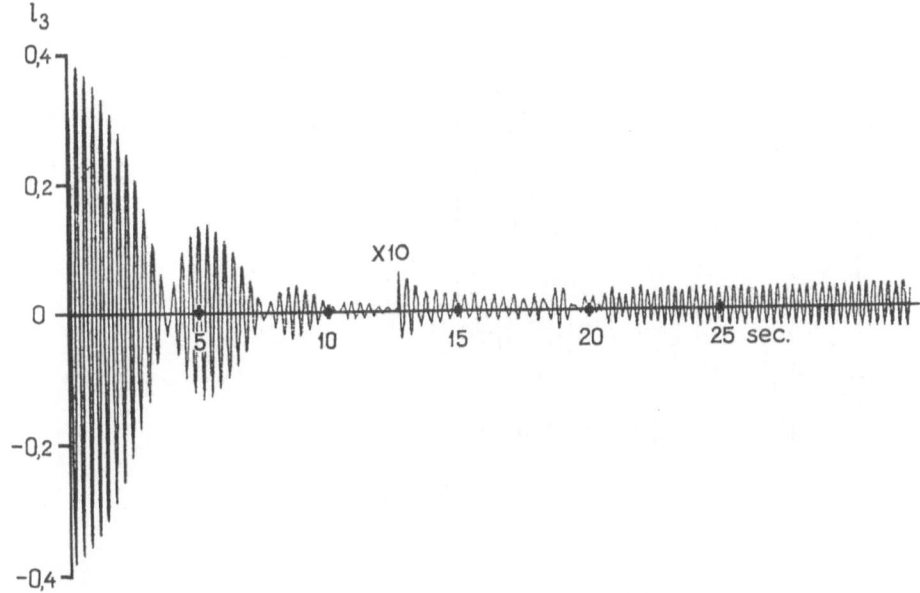

Fig. 3. Typical Differential Analyzer Solution for the Spin-Modulated Control System.

the output of the rate gyro, we will obtain a pulse which can be used to open the control-jet. If a single control-jet is used in the nose of the rocket and is aligned say, along the x-axis of the rocket, then L_y can only have a positive sign. This means that an erecting-torque impulse based on the null in bank angle Φ will occur only once each revolution, while a damping-torque impulse based on the null in ω_x will also occur only once each revolution. Thus only pulses of one sign coming from the bank angle indicator (free gyro or horizon scanner) will open the control-jet, and similarly for the ω_x rate gyro, only output pulses of one sign are used.

Electronic Differential Analyzer Simulation of the Control System

To study the spin-modulated attitude-control system on the electronic differential analyzer, a particular rocket combination was selected, namely a Recruit (loaded weight equals about 370 lb) followed by an Asp (loaded weight equals about 190 lb) plus a 30-lb payload. The single control-jet was located in the payload section. A spin rate of 3 rps was chosen. If an air bottle is used as the gas source for the control-jet, the thrust will fall off roughly exponentially with time. For this reason we assumed such a fall off in available erecting torque L_y, with a time constant of 10 seconds.

A typical computer solution starting with the rocket 20° off the vertical is shown in Fig. 3. Plotted is $l_3 = -\sin \Theta \approx -\Theta$. As explained earlier, Θ oscillates at ω_z, the rocket spin frequency of 3 rps. The envelope of the Θ-curve actually represents the angle α between the rocket longitudinal axis and the vertical. For the example shown, a total control-jet impulse of 275 lb sec is required. For a 20% duty cycle (the jet is fired over a 72° sector), this represents an initial thrust level of about 83 lb for the erecting impulses, $83/1.67 = 50$ lb for the damping impulses. As stated in the previous paragraph, these impulses fall off by a factor of $1/e$ or 37% every

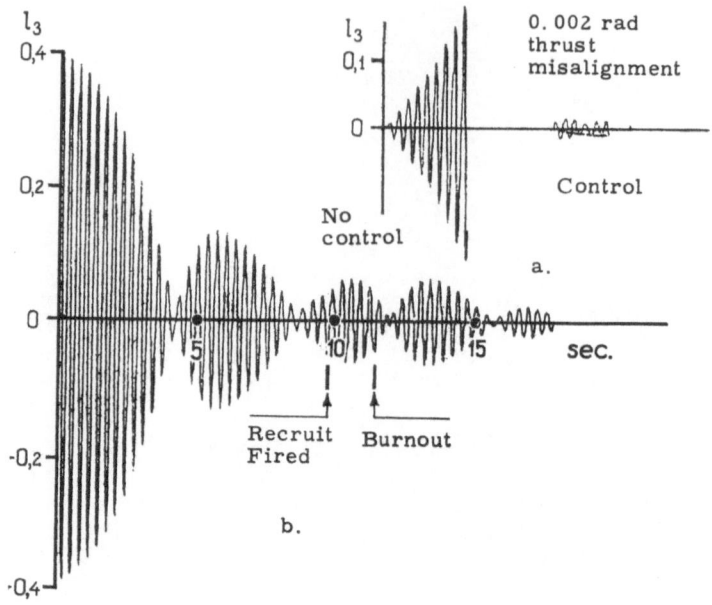

Fig. 4. Effectiveness of Spin-Modulated Control System in Offsetting Thrust Misalignment (0.002 Radian Misalignment Assumed).

10 seconds. Note that an oscillation of somewhat steady amplitude finally results; this amplitude is about 0.2°.

The effectiveness of the control system when the Recruit rocket is fired with a 2-mil thrust misalignment is shown in Fig. 4. In Fig. 4 'a' is the result for an initial rocket attitude angle of zero, while in 4 'b' is the result when the initial angle is 20°.

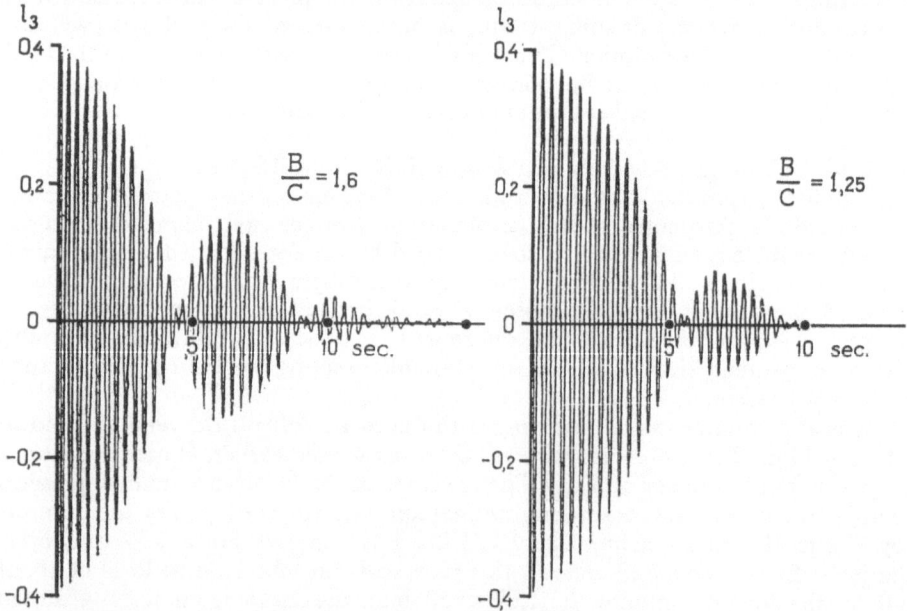

Fig. 5. Effect of Different Erecting-to-Damping Torque Ratios B/C.

94

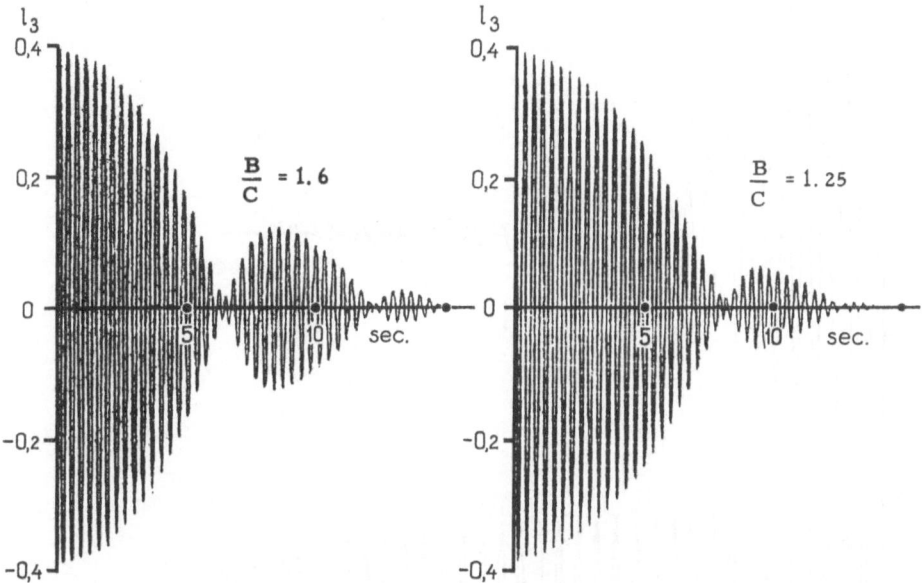

Fig. 6. Effect of Factor of Two Decrease in Thrust Level Compared with Figure 5.

In both cases the thrust of the control jet is ½ that used for the previous figure. Thus total impulse required is about 140 lb sec, which could be obtained with approximately 2 lb of air (specific impulse $\approx$ 70 sec). It should be noted that no falloff in inertia with fuel consumption was represented in these curves, nor was the jet damping effect included.

In Fig. 5 the effect of two different erecting-torque-to-damping-torque-ratios B/C is shown; the first curve shows the transient for $B/C = 1.6$, while the second curve shows the result for $B/C = 1.25$. Fig. 6 shows the effect of decreasing the thrust level by a factor of two but at the same time increasing the time constant of the air-bottle thrust-decay curve from 10 to 20 seconds.

The rate-gyro sensitivities required for this damping technique are high, and it is important that the rate gyro axis of sensitivity be perpendicular to the spin axis of the rocket. Otherwise a component of the spin velocity ω_z will be measured by the gyro with the result that a larger limit cycle is reached. This does not mean that the overall accuracy of the attitude control is decreased appreciably, however.

The damping signal can also be obtained from a conventional accelerometer mounted on the z-axis of the rocket with its axis of sensitivity along the x-axis. Since the rocket is already spinning about the z-axis the accelerometer will measure a centrifugal acceleration dependent on ω_x. Null output of this signal can be used to activate the damping command.

The Recruit-Asp rocket combination used as an example here has a very low I_x/I_z ratio, approximately 1/700. This is because both rockets are long and slender. A rocket such as the third-stage Vanguard has an I_x/I_z ratio of about 0.2 when loaded with propellant. Here the angular momentum due to the spin ω_z is much more important and there will be a strong tendency for the rocket to precess rather than erect following a control-jet impulse. This is evident in Fig. 7, where computer solutions for $I_x/I_z = 0.02$ and 0.2 are shown. In the latter case the spin angular momentum ruins the erection process. However, the effect decreases as $1/\omega_z^2$, so

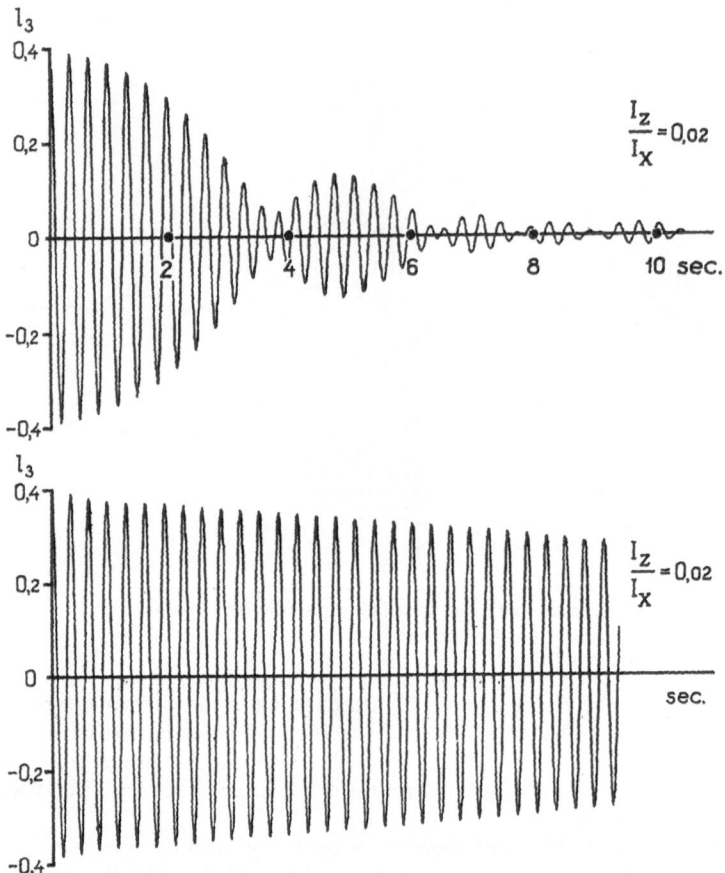

Fig. 7. Effect of Different Inertia Ratios I_z/I_x.

that by using a spin-frequency less than the 3 rps considered, one could still presumably employ the spin-modulated control scheme. Since short, fat rockets tend to burn longer, a lower spin frequency should be practical.

Possible Gas Sources for the Jet Controls

We have already discussed the use of compressed air as a gas source for the control-jet. For the Recruit-Asp combination used in the previous section, about 2 lb of air was considered. This requires about an 8-in. sphere at 4000 psi, which seems not unreasonable. Total weight including the container might be approximately 10 lb. For a smaller rocket, the size and weight could be reduced accordingly.

As long as the bottle weight is larger than the gas weight, it is profitable to use heavier gases even though their specific impulse is less. Another factor which one must consider is the cooling of the gas due to expansion. This will probably cause some thrust loss due to liquifaction as well as impulse decrease resulting from the lower temperature.

To decrease the required thrust level and simplify valving, a two-nozzle system can be used, the nozzles being on opposite sides of the rocket. In this scheme the gas source would never be turned off, and the valve is only required to deflect the air through one nozzle or the other. In this case the effective impulse in the desired direction integ ated over a half revolution is actually $2/\pi$ times the impulse if it were all concenrrated in the desired direction. Thus the erecting impulse when the nozzle is left open for a half revolution is 63% effective, which is probably a worthwhile price to ray for the simplicity and lower thrust level. By opening the nozzle 90° ahead of the instant when it will be oriented to give maximum erecting impulse and closing it 90° after this instant, the timing of the spin-modulated control system can be achieved. Over the following half revolution, the opposite jet is opened. Damping is achieved by opening one or the other nozzle over some fraction of a half revolution. When the damping signal has the same sign as the erecting signal, the appropriate nozzle remains open over the full half revolution; when the damping signal has the opposite sign from the erecting signal, the appropriate nozzle opens during the early portion of the half revolution, is then closed, and the other nozzle opened by the damping signal; and during the last part of the half revolution the first nozzle is opened again. The net result is a lower net erecting torque. Study of this scheme with the electronic differential analyzer has indicated that it works perfectly satisfactorily.

Action times of valves considered suitable for controlling the air flow are in the order of 10 milliseconds, which appears satisfactory for spin rates as high as 3 rps.

One nice feature of the compressed air system is the thrust decay with time. The maximum thrust level occurs initially, when it is most needed to erect the rocket as quickly as possible from its initial error in attitude. During the firing of the lower stage, the thrust required to hold the attitude becomes less, and during the upper-stage firing, the thrust needed becomes still less. After burnout the thrust needed to hold the payload attitude in free fall is even smaller. Thus the spin-modulated attitude-control system using an air bottle should continue to be effective during a large part of the free fall. To lengthen this time, one could conceivably switch to smaller nozzles after burnout.

We have also looked into hydrogen peroxide as a gas source. The Johns Hopkins Applied Physics Laboratory has developed such a system for reaction controls. It has the advantage that one can use a valve to control the liquid H_2O_2 instead of hot gases. By passing the liquid through a catalyst in the nozzle throat, the reaction is started and the hot gas expands through the nozzle. The liquid peroxide can be contained in a bottle pressurized to several hundred psi by nitrogen. This low pressure should allow lighter bottle weight than in the case of air. Also, H_2O_2 has a higher specific impulse, around 150. It has been estimated that a total impulse of several hundred lb sec could probably be obtained with a system weighing 5—8 lb, including bottle, valve, plumbing, and nozzle. The disadvantage to H_2O_2 is that it requires somewhat special handling.

Another source of hot gas which should be considered is end-burning solid-fuel sticks. Such fuel elements have received considerable attention as a source of gas for auxiliary-power in missile control surface actuators. Typical specific impulse is about 180. Burning rates vary, depending on the type of fuel, but 1/8 in/sec is representative. For the thrust level needed on a Recruit-size rocket, a fuel-rod diameter of several inches will be required. Weight of the plastic inhibitor case and steel jacket should be comparable to the fuel weight, so that a total specific impulse of about 100 seems possible. If the total impulse required is only several hundred, a very light gas source is obviously present in the form of a solid-fuel rod and case.

The big disadvantage of the solid-fuel gas source is the high temperature. However, flapper-type valves have been developed which can handle the temperatures for long periods (minutes) of time. Some insulation of the gas source from the rest of the rocket instrumentation would, of course, be necessary. The temperatures are not nearly as high as those for high-performance propellants. Chamber pressures of the order of 500 psi allow fairly light steel tubing to be used for the case.

Obviously a hot gas source such as this would have to be ejected continuously, so that the scheme described at the beginning of this section would be used, *i.e.*, two diametrically opposite nozzles with first one opened and then the other. By tapering the solid-fuel stick or by reducing the diameter in discrete jumps, we could probably reduce the mass flow and hence the thrust level as the flight progressed.

It is felt that all three of the above gas sources for control-jet reaction should still be considered. There appears to be no clear choice at the present time, although, as hardware details are worked out one will undoubtedly appear more attractive than the other.

Method of Generating the Required Spin Rates

It is apparent that the spin-modulation method of attitude control requires that the rocket spin in a predetermined direction. Further study is needed to determine the tolerance in spin rate which can be allowed and still maintain control effectiveness, but it is anticipated that this tolerance will be considerable (*e.g.*, 3 rps $\pm$ 2 rps). Obviously one would like to use a completely passive system such as canted fins on the boost stage to produce this spin. However, experience with previous sounding rockets indicates that this possibility is questionable.

The next simplest technique might be to use a despinning device such as the unfurling of weights on cords. This would slow the spin of the upper stages down to a negligible value once outside the atmosphere. Small spin rockets could then be fired to spin up the assembly to the desired rate.

A third method would be a full closed-loop system consisting of a source of gas (*e.g.*, compressed air) with nozzles oriented to give spin moments in either direction. A rate gyro, or possibly an accelerometer at the perimeter of the rocket, would measure the spin rate, and depending on the sign of the rate error, a flapper valve would open one pair of nozzles or the other. Since this is a rate-control system, no special damping methods need to be employed. It is fundamentally a simple bang-bang servo. For a Recruit-size rocket, a thrust level of only 1 lb each nozzle would spin the rocket to 1 rps in about 3 seconds. If in the payload section, this device could be used to despin the payload later during the flight. It could also be located off the upper stages in a section between the boost and upper stages. This section would also support any aerodynamic fins needed to stabilize the upper stages after booster separation and during the coast out of the atmosphere. Upon upper-stage ignition, it would be left behind, thus eliminating the weight penalty of fins and spin-up system.

Conclusion

It is concluded that the spin-modulation method of attitude control allows control of pitch and yaw angle using a single auxiliary control-jet, or at most, two back-to-back jets. The system is useful for attitude control during both thrusting and coasting portions of flight. If an accelerometer is used to provide damping signals and a horizon scanner as an attitude reference for vertical orientation, then a complete autopilot system for attitude control of a 500—1000 lb rocket would need only one moving part (the control-jet valve) and should weigh 5—10 lb. Such a system would be useful for vertical sounding rockets, space probes, and lunar soft-landing vehicles.

The author would like to acknowledge the financial support of the Ballistics Research Laboratories of the US Army for carrying out the investigation reported in this paper.

Direct Measurements of Ion Density and Conductivity in the D-region

E. C. Whipple, Jr.

Goddard Space Flight Center, National Aeronautics and Space Administration, Washington 25, D.C.

Certain low altitude atmospheric electricity techniques can in principle be extended to higher altitudes. An ion collector adapted for rocket determination of ionic density, conductivity and mobility in the D-region is described. Previous results point out the importance in general of dust in providing a recombination surface for ions and electrons.

The electrical properties of the D-region are perhaps less understood than any other region of the ionosphere below the exosphere. This is particularly true of the normal behavior of the D-region as opposed to black-out occasions where there is abnormal ionization. The reasons for this are the low ion and electron densities and their high collision frequencies which are disadvantageous to the usual radio-sounding techniques.

On the other hand, the electrical properties of the lower atmosphere — below about 30 km — have been the objects of investigation for many years by direct measurement techniques. Ion densities in the D-region are relatively high when compared with the troposphere where they are measured in hundreds rather than in thousands. In principle the D-region is susceptible to investigation by many of these direct measurement techniques, and the advent of the sounding rocket has overcome the problem of inaccessibility.

This paper describes an ion collector experiment adapted for rocket measurement of ionic density and conductivity in the D-region, and comments on some experimental results from the Viking 5 rocket which have been reported in detail elsewhere [1].

Ion Collector Instrumentation

The Gerdien condenser is an ion collector consisting of two coaxial cylindrical electrodes which is useful for measuring atmospheric conductivity. The usual analysis of the instrument's operation is valid as long as the ionic mean free path is small compared with the dimensions of the condenser, that is, up to about 80 km. Two condensers mounted on the nose of a rocket are shown in Fig. 1. As air flows through the cylinder the ions are separated by an applied electric field between the electrodes. Ions of one sign are collected by the center electrode and the current is measured as a function of the potential difference between the electrodes. An ideal current-voltage characteristic is shown in Fig. 2. The slope of the linear portion of the curve through the origin is proportional to the atmospheric conductivity, and the saturation current is proportional to the ion density. The voltage at the knee of the curve is inversely proportional to the ionic mobility.

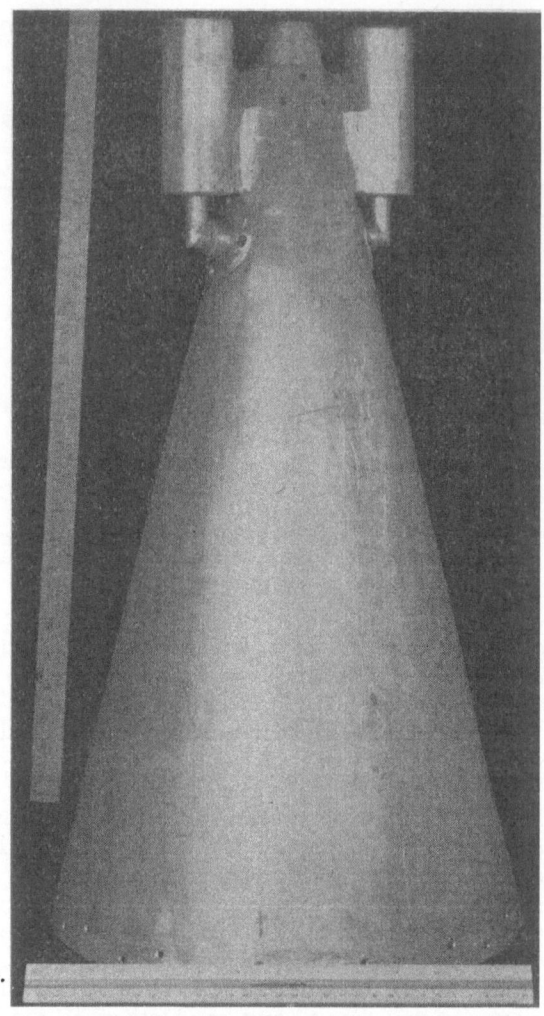

Fig. 1. Rocket Mounted Ion Collectors.

Possible effects from such things as a charge on the rocket or from the rocket's shock wave have been evaluated elsewhere [1]. When this experiment was flown on the Viking 5 the collecting potential was programmed in 12.5 volt steps between —100 and +100 volts. It would be desirable to obtain better resolution of the current-voltage characteristic. However, the time to obtain one curve must be held to about one second or less so that the atmospheric quantities to be measured remain relatively constant over the corresponding altitude interval through which the rocket moves. An externally applied collecting potential cannot be swept continuously because the resultant charging current to the collector would mask the ion current which is much smaller. Thus, the input resistor must be short-circuited while the applied potential is changed, and an increase in the number of voltage steps per second begins to demand better response times of low-current electrometers than have so far been obtainable.

In Fig. 3 a scheme is shown whereby the ion current itself can be used to sweep the collecting potential continuously without introducing undesirable effects. Initi-

Direct Measurements of Ion Density and Conductivity in the D-region

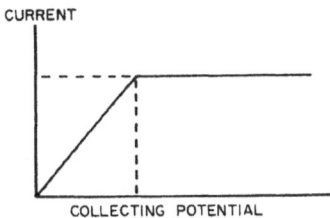

Fig. 2. Ideal Current-Voltage Characteristic.

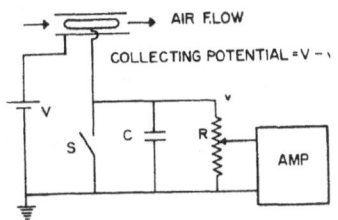

Fig. 3. Ion Collector Input Circuit.

ally, with the switch S closed, the collector is at ground potential and the collecting potential is equal to the battery potential V. When the switch is opened, the collector potential rises at a rate determined by the circuit parameters and the ion current being collected. By measuring both the collector potential and its rate of change the current-voltage characteristic can be determined, the resolution essentially being limited only by the telemetry response. A new characteristic curve will be obtained each time the switch is actuated.

It is planned to fly this experiment in an Aerobee Jr. at Fort Churchill in October of this year. Atmospheric pressure will be measured simultaneously.

Results

One unexpected result of the Viking 5 experiment was that above 50 km the observed conductivity was lower than that expected from ionization by cosmic rays alone. This was attributed to ionic diffusion to dust, the reduction in conductivity agreeing quantitatively with that calculated from estimates of the radius and concentration of noctilucent cloud particles. The importance in general of dust in providing a recombination surface for ions and electrons in the upper atmosphere does not seem to have been sufficiently recognized in the past. There are two areas where this effect may be important.

In the D-region the rate of loss of positive ions is determined by recombination with electrons and negative ions and possibly by diffusion to dust particles. The expressions for these reactions are shown in Table I. (N is the appropriate particle density, α the recombination coefficient, a the dust particle radius, and c_+ the mean thermal velocity of the positive ions.) The numberical values shown for a and N_0 are not particularly conservative but are within the range of estimates which have been published [3]. It is apparent that diffusion to dust predominates over recombination with electrons as long as the electron density is less than about 500/cm³, which is the case for most of the normal D-layer region.

Diffusion to dust will also predominate over ion-ion recombination whenever the negative ion density is below a certain value dependent upon altitude as shown in Fig. 4. Also shown are the negative ion densities calculated from the Viking 5 conductivity results. Again, diffusion to dust may well be dominant above 50 km.

The other area where the role of dust may be important is in interplanetary space, as has been suggested by Shklovskii [4]. Electrons are lost by recombination with ions (protons) and again possibly by diffusion to dust particles as shown in

TABLE I. Rate of Loss of Positive Ions in the D-Region

$$\text{Rate} = \underset{\text{A}}{\alpha_e N_e N_+} + \underset{\text{B}}{\alpha_i N_+ N_-} + \underset{\text{C}}{\pi a^2 c_+ N_+ N_0} \qquad \text{C/A} > 1 \text{ for } N_e < 500/\text{cm}^3$$

$$\text{where} \quad \alpha_e = 3 \cdot 10^{-8} \text{ cm}^3/\text{sec} \ [2] \qquad a = 10^{-5} \text{ cm,}$$
$$c_+ = 5 \cdot 10^4 \text{ cm/sec } (300\,°\text{A}), N_0 = 1/\text{cm}^3.$$

TABLE II. *Rate of Loss of Electrons in Interplanetary Space*

$$\text{Rate} = \underset{D}{\alpha_H N_e^2} + \underset{E}{\pi a^2 c_- N_0 N_e} \qquad E/D > 1 \text{ for } N_e < 30/cm^3$$

where $\quad \alpha_H = 2 \cdot 10^{-14}$ cm³/sec $\qquad a = 3$ microns
$c_- = 6 \cdot 10^7$ cm/sec (10^4 °A), $N_0 = 3 \cdot 10^{-14}/cm^3$

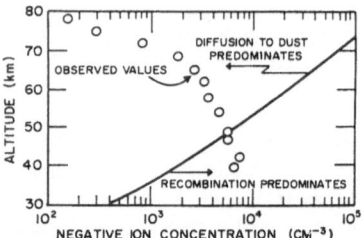

Fig. 4. Dependence of Positive Ion Loss
on Negative Ion Concentration.

Table II. Estimates of the hydrogen recombination coefficient vary considerably. Using a value for α of $2 \cdot 10^{-14}$ cm³sec estimated from work by Burgess [5], and values of $3 \cdot 10^{-14}/cm^3$ and 3 microns for the dust concentration and radius as indicated by micrometeorite flux measurements from rockets and satellites [6] it appears that this effect may be important whenever the electron concentration is less than about 30/cm³. Blackwell [7] has recently estimated from zodiacal light observations that N_e is probably less than 50/cm³ in the vicinity of the earth's orbit.

In conclusion, two points are emphasized: first, the ion collector familiar in low altitude atmospheric electricity research is considered to be useful for D-layer investigation. Possibly other low altitude techniques could be adapted for rocket use in this region. Finally, the recombination surface for ions and electrons that is provided by dust may well be important in the D-layer and in interplanetary space.

References
[1] *Bourdeau, R. E., Whipple, E. C. Jr.,* and *Clark, J. F.,* J. Geophys. Res., *64,* 1363 (1959).
[2] *Bailey, D. K.,* Proc. I. R. E., *47,* No. 2, 255 (1959).
[3] *Ludlam, F. H.,* Tellus, *9,* 341 (1957).
[4] *Shklovskii, I. S.,* Soviet Astronomy — AJ, *2,* 516 (1958).
[5] *Burgess, A.,* M. N., *118,* 477 (1958).
[6] Geophysics Research Directorate, Research Note No. 28, Air Force Cambridge Research Center, (1960). (AFCRC-TN-59-647)
[7] *Blackwell, D. E.,* Presented at the National Academy of Sciences — American Geophysical Union Symposium, April, 1960, Washington, D.C.

Problems of Aeronomic Chemical Reactions

Marcel Nicolet

Centre National de Recherches de l'Espace, Brussels, Belgium

In view of the fact that, in the neighbourhood of the mesopause, there are differences between mesopheric and thermosperic composition, it is necessary to study atmospheric constitution by means of small rockets.

The peak of the atomic oxygen concentration should be determined in the lower thermosphere. The exact ratio of molecular oxygen and nitrogen is an important parameter to be found. Information regarding latitudinal difference should be obtained. The problem of nitrogen oxides is also an important problem of the lower thermosphere.